Preventing and Managing Disabling Injury at Work

Preventing and Managing Disabling Injury at Work

Edited by
Terrence Sullivan and John Frank

Taylor & Francis
Taylor & Francis Group

LONDON AND NEW YORK

First published 2003 by Taylor & Francis
11 New Fetter Lane, London EC4P 4EE

Simultaneously published in the USA and Canada
by Taylor & Francis Inc.,
29 West 35th Street, New York, NY 10001

Taylor & Francis is an imprint of the Taylor & Francis Group

Publisher's note:
This book was prepared from camera-ready-copy supplied by the authors/editors.

Printed and bound in Great Britain by The Cromwell Press, Trowbridge, Wiltshire.

Every effort has been made to ensure that the advice and information in this book are true
and accurate at the time of going to press. However, neither the publisher nor the authors
can accept any legal responsibility or liability for any errors or omissions that may be made.
In the case of drug administration, any medical procedure or the use of technical equipment
mentioned within this book, you are strongly advised to consult the manufacturer's guidelines.

British Library Cataloguing in Publication Data
A catalogue record for this book is available from the British Library.

Library of Congress Cataloging in Publication Data
Preventing and managing disabling injury at work/edited by Terrence Sullivan and John Frank.
    p. cm.
    Includes bibliographical references and index.
    ISBN 0-415-27491-5 (hb.)
    1. Industrial safety. 2. Accidents—Prevention. I. Title: Preventing and managing disabling
    injury at work. II. Sullivan, Terrence James, 1951– III. Frank, John.
T55 .P745 2003
658.3'82—dc21

                        2002032105

ISBN 0-415-27491-5

"Science may not have the answers that some businesses, employers, or policy-makers want to hear. But the time has come to listen to what is known about MSDs (musculoskeletal disorders) – and to use that knowledge to make wise decisions in the years to come."

Jeremiah A. Barondess, President of the New York Academy of Medicine, New York City, Chair of the National Research Council/Institute of Medicine committee which produced the report Musculoskeletal Disorders and the Workplace: Lower Back and Upper Extremities. *National Academies News Release, Feb. 9, 2001.*

In memory of Terry Thomason (1950-2002) who led the way in Canada-US comparisons of workers' compensation systems.

Contents

THEME 3: Stakeholder Engagement in Intervention Programs

THEME 4: From Evidence to Policy and Practice

Contributors

Rachelle Buchbinder is Director of the Department of Clinical Epidemiology at Cabrini Hospital and Associate Professor in the Department of Epidemiology and Preventive Medicine at Monash University in Melbourne, Australia. A rheumatologist, her research interests include treatments for common regional musculoskeletal disorders and estimating the risk of malignancy in rheumatic diseases. She was co-recipient, with Damien Jolley and Mary Wyatt, of the Volvo Award in Clinical Sciences in 2001. E-mail: rachelle.buchbinder@ med.monash.edu.au

Donald Cole completed his MD at the University of Toronto in 1978 and his MSc in Design, Measurement and Evaluation in Health Services at McMaster University in 1991. Cole is a Royal College specialist in Occupational Medicine and in Community Medicine. He is an Associate Professor in the Department of Public Health Sciences at the University of Toronto and Senior Scientist at the Institute for Work & Health. He is an expert in work-related musculoskeletal disorders of the upper extremities. He holds a number of grants and is active in international health work. E-mail: dcole@iwh.on.ca

Marie-José Durand is an occupational therapist and assistant professor at the University of Sherbrooke (Department of Community Health Sciences). Her doctoral and postdoctoral studies focused on program evaluation in the field of work rehabilitation for musculoskeletal disorders. Her current work is dedicated to the development and assessment of conceptual models and tools for return to work programs. E-mail: marie-jose.durand@courrier.usherb.ca

Renée-Louise Franche is a scientist at the Institute for Work & Health in Toronto, Canada. Her research focuses on understanding the contribution of organizational, individual and healthcare provider factors in determining safe and sustainable return to work following an injury or illness. She is an Assistant Professor in the Department of Public Health Sciences and the Department of Psychiatry, Faculty of Medicine, University of Toronto. She is an associate member of the University Health Network Women's Health Program. E-mail: rfranche@iwh.on.ca

John Frank received his MD from the University of Toronto and completed his MSc in Community Health in Developing Countries at the London School of Hygiene and Tropical Medicine. In 2001, Dr Frank was appointed Scientific Director of the Institute of Population and Public Health at the Canadian Institutes of Health Research. He was a visiting scholar at the School of Public Health, University of California at Berkeley, and is currently a Professor in the Department of Public Health Sciences, Faculty of Medicine, University of Toronto. He is also Senior Scientist and former Research Director of the Institute for Work & Health, and his research interests include work and health and social epidemiology. He is also a fellow in the population health program for the Canadian Institute for Advanced Research. E-mail: john.frank@utoronto.ca

Gary M. Franklin, MD, MPH, is the Medical Director of the State of Washington Department of Labor and Industries in Olympia, Washington USA. He is also a Research Professor, Department of Environmental Health and Medicine, University of Washington, in Seattle, Washington. Dr Franklin obtained an MD degree from George Washington University, and a MPH degree from the University of California at Berkeley. A board-certified neurologist, his current positions also include being Medical Director, Washington State Department of Labor and Industries. His major research interests include use of administrative databases (workers' compensation) to conduct epidemiologic research, outcome of medical treatment modalities for occupational musculoskeletal disorders, predictors of disability in workers' compensation, and impact of managed care delivery systems on cost, outcome and satisfaction in workers' compensation. E-mail: fral235@lni.wa.gov

Jaime Guzman obtained his MD degree and specialties in Internal Medicine and Rheumatology in Mexico. He obtained his MSc in Clinical Epidemiology at the University of Toronto and is currently completing training in Physical Medicine and Rehabilitation in the Faculty of Medicine, Winnipeg, Canada. His current research interests are in disability prevention in musculoskeletal conditions, back pain in particular. He is a Research Associate with the Occupational and Environmental Health Unit, Department of Community Health Sciences, University of Manitoba. E-mail: guzmanj0@cc.umanitoba.ca

Damien Jolley is Associate Professor of Biostatistics and Epidemiology in the School of Health Sciences at Deakin University, Australia. His career in the methodology of public health research has extended from malaria and nutrition in Papua New Guinea, cancer and small area health statistics in Australia and the United Kingdom and more recently physical activity and mass media campaigns. He was joint winner, with Rachelle Buchbinder and Mary Wyatt, of the Volvo Award in Clinical Sciences in 2001. E-mail: damien@jolley.com.au

Mickey Kerr holds appointments as an Assistant Professor in the School of Nursing, Faculty of Health Sciences, at the University of Western Ontario, and as an Adjunct Assistant Professor in the Department of Public Health Sciences, Faculty of Medicine at the University of Toronto. He is also a Scientist in the Workplace Studies group at the Institute for Work and Health in Toronto. His research interests include risk factors for musculoskeletal disorders, the workplace psychosocial environment, stress research, and biological mechanisms for injury.

Much of his recent work has been focused on the work environment and health in the health sector. E-mail: mkerr@iwh.on.ca

Niklas Krause received his medical degree (MD) and a doctoral degree (Dr med.) in Orthopedic Medicine from the University of Hamburg, Germany, and specialized in orthopedic rehabilitation (physiatry). In the United States he earned a master's degree in public health and a doctoral degree (PhD) in epidemiology from the School of Public Health at the University of California, Berkeley. His research focus has been the epidemiology and prevention of work-related musculoskeletal diseases and disability. He currently holds a position as Assistant Professor of Medicine in the Division of Occupational and Environmental Medicine of the University of California at San Francisco. E-mail: nkrause@uclink4.berkelcy.edu

Patrick Loisel is an orthopedic surgeon at Charles LeMoyne teaching hospital and full professor at the University of Sherbrooke. His work is dedicated to research and research transfer in the field of work disability prevention and work rehabilitation for musculoskeletal disorders. He has developed, tested and applied in the community an evidence-based, work disability management program, linking clinical and workplace interventions. He is the leader of a multidisciplinary clinical and research team in work rehabilitation funded by the Quebec Health Research Grant Agency (FRSQ) and the director/founder of the Quebec Network in Work Rehabilitation (RRTQ). E-mail: ploisel@courrier.usherb.ca

John Mendeloff is a Professor at the University of Pittsburgh with appointments in the Schools of Public and International Affairs, Public Health, and Law. He has written extensively on American public policies on occupational safety and health, including Regulating Safety and The Dilemma of Toxic Substance Regulation. He has served on many government commissions reviewing policy on this topic and has often consulted for OSHA. He has recently carried out extensive work evaluating OSHA's enforcement and consultation programs. E-mail: jmen@birch.gspia.pitt.edu

Robert D. Mootz is the Associate Medical Director for Chiropractic in the State of Washington Department of Labor and Industries, Olympia, Washington, USA. Dr Robert D. Mootz holds a degree in chiropractic and biology from Palmer College of Chiropractic and the University of California. His research and health policy interests include quality improvement, evidence based clinical practice, and occupational health outcomes. He is the Editor of Topics in Clinical Chiropractic and has authored texts on chiropractic. He co-edited the US Government Report, Chiropractic in the United States – Training, Practice and Research. In addition to 13 years in private practice, Dr Mootz also served as a Professor at Palmer College of Chiropractic West and has served in leadership roles in the American Back Society, American Public Health Association and is adjunct faculty for three chiropractic colleges. E-mail: moot235@lni.wa.gov

Robert Norman is Professor Emeritus of occupational biomechanics and ergonomics in the Department of Kinesiology and former Dean of the Faculty of Applied Health Sciences at the University of Waterloo, Canada. He chairs the Research Advisory Council of the Workplace Safety and Insurance Board of Ontario and is a former Volvo Award recipient for biomechanics research on the lumbar spine. He has worked on both laboratory and industrial field research on quantifying low back injury risk for many years and has consulted extensively on ergonomics issues for business, industry and governments. His most recent research has been identification of the relative contributions of biomechanical and psychosocial risk factors to the reporting of low back pain in a large auto assembly plant. This has been complemented by current studies of the effectiveness of participatory ergonomics interventions aimed at preventing injury while maintaining quality and productivity. E-mail: norman@healthy.uwaterloo.ca

Aleck Ostry is the recipient of a Canadian Institute for Health Research scholar award and is currently an Assistant Professor in the Department of Healthcare and Epidemiology and the Centre for Health Services and Policy Research at the University of British Columbia (UBC). He has an MSc in Health Service Planning from UBC, an MA in history from Simon Fraser University (specializing in the history of public health), and a PhD in epidemiology (UBC). He teaches courses on the social determinants of health and conducts research on work stress, and the impacts of change in work organization, labour markets, and technology on health. E-mail: ostry@interchange.ubc.ca

Michael Polanyi is Assistant Professor in the Department of Kinesiology and Health Services at the University of Regina. He was an Associate Scientist at the Institute for Work & Health in Toronto, where he studied the ways in which workplace characteristics, work experiences and socio-economic conditions affect health. He is also interested in collaborative processes through which researchers can contribute to social change. He recently completed a two-year participatory research project led by injured workers to better understand and to improve the compensation and return-to-work system in Ontario. He holds a doctorate in Environmental Studies, obtained from York University in 1999. He has worked professionally in the areas of adult education, community development and health promotion. E-mail: michael.polanyi@uregina.ca

Jerry Spiegel is an Assistant Professor at the Liu Centre for the Study of Global Issues and the Institute of Health Promotion Research at the University of British Columbia. Between 1985 and 2000 he served as Director of Planning Research and Evaluation for the Manitoba Government Department of Environment and Workplace Safety and Health, Director of Pollution Prevention for Manitoba Environment and Director of Regulation Review with the Workers Compensation Board of British Columbia where he maintained a particular research interest in the evaluation of population health prevention-oriented programs. Dr Spiegel received his PhD from the University of Manitoba, an MSc from the University of Toronto, an MA from McGill University and a Diploma in Occupational Health and Safety from McMaster University. E-mail: jerrymspiegel@aol.com

Terrence Sullivan is Vice President, Preventive Oncology, at Cancer Care Ontario. Previously the president of the Institute for Work & Health in Toronto, he also played an active role as workplace leader in HealNet, a government of Canada sponsored center of excellence in health research. He is Associate Professor in the Departments of Health Policy, Management and Evaluation and Public Health Sciences at the University of Toronto. He was a senior Ontario official in both Health Policy and Intergovernmental Relations, and is interested in comparative approaches to health and disability policy. His recent books include the previous collection *Injury and the New World of Work* (University of British Columbia Press, 2000) and *First Do No Harm; Making Sense of Canadian Health Reform* (with P. Baranek, Malcolm Lester Books, 2002). E-mail: terry.sullivan@cancercare.on.ca

Terry Thomason (deceased, April 2002) was appointed Director of the Charles T. Schmidt, Jr. Labor Research Center at the University of Rhode Island in 1999. He had previously taught for eleven years at the Faculty of Management of McGill University in Montreal, Canada. He received his PhD from the New York State School of Industrial and Labor Relations at Cornell University in 1988. His work has been published in a wide range of professional journals, including Industrial and Labor Relations Review, Journal of Legal Studies, Journal of Risk and Insurance, and Journal of Labor Economics. He was author (with Timothy P. Schmidle and John F. Burton, Jr.) of *Workers' Compensation: Benefits, Costs, and Safety Under Alternative Insurance Arrangements*, published by the Upjohn Institute, and editor of *New Approaches to Disability in the Workplace* (with John F. Burton, Jr. and Douglas E. Hyatt).

Thomas Wickizer has applied his training in health economics and policy to address questions related to utilization management, managed care, health care expenditure growth, employer health insurance costs, substance abuse treatment, and workers' compensation. In November 1997, Dr Wickizer was awarded the Rohm & Haas Distinguished Professorship to pursue research activities in the area of occupational health care. Dr Wickizer's current research includes: an evaluation of the effects of workplace drug-free programs on occupational injuries and medical costs; evaluation of a multi-year quality improvement project for the Washington State workers' compensation program; analysis of employment outcomes among AFDC recipients treated for substance abuse in Washington State; and a follow-up study of long-term medical outcomes among injured workers treated in managed care and traditional fee-for-service arrangements. Dr Wickizer is Program Director of a five-year training program in occupational health services research at the University of Washington, School of Public Health. E-mail: tomwick@u.washington.edu

Mary Wyatt has practiced in Occupational Medicine for 15 years and has worked extensively on return to work strategies. She has spent time mediating in workers' compensation disputes and has visited more than a thousand work sites to explore return to work tasks and issues. Mary lectures in Occupational Medicine at Monash University, has sat on a number of government committees dealing with work related problems, has been involved in research into changing back pain beliefs, the development of guidelines for management of work-related problems for treating practitioners, and is active in clinical practice. She is part of a company that works with employers to develop effective management of health and safety, injuries and return to work. E-mail: mwyatt@ozemail.com.au

Annalee Yassi, an occupational physician and epidemiologist, is currently the Director of the Institute of Health Promotion Research (IHPR) at the University of British Columbia, and is the Founding Executive Director of the Occupational Health and Safety Agency for Healthcare in BC (OHSAH). She also holds a Canada Research Chair, has served on numerous national and international commissions, and has published extensively. Her special interest is the link between the workplace and clinical treatment, and the collaborative roles that must be played by the various stakeholders in injury prevention and disability management. As Director of the Department of Occupational and Environmental Medicine at Winnipeg's Health Sciences Centre for more than 12 years, she built a multidisciplinary comprehensive occupational health program for healthcare workers that has been highly successful. Dr Yassi received her Medical Degree from McMaster University, her Masters Degree from the University of Toronto and she is a Royal College Fellow in Community Medicine and Occupational Medicine. E-mail: annaleey@aol.com

Acknowledgements

This book arose from the labor of many hands. We would like to express our thanks first and foremost to our collaborators in HealNet, a network of centers of excellence funded by the government of Canada. HealNet has supported each of us in our efforts and, more importantly, supported the coming together of a group of talented researchers, many of whom have made contributions in this book. In addition to our collaborators, the network leadership within HealNet – George Browman, Vivek Goel and Diana Royce – has been very active in their support for the unique kind of workplace health research which is covered in this book. Our efforts within the network are here complemented with able collaborators and friends from the USA (Terry Thomason, John Mendeloff, Gary Franklin, Robert Mootz and Thomas Wickizer) and from Australia (Rachelle Buchbinder, Damien Jolley, Mary Wyatt) who, in the context of different anglo-policy environments, share many common challenges with us in the struggle to understand, prevent and manage modern workplace disabilities. For each of us, as editors, the support of the Institute for Work & Health, its staff and Board have been essential to supporting the workplace research within HealNet and to making this book a reality. Vincy Perri was especially helpful in pulling together the early elements of this project. In addition, thanks are due to Cancer Care Ontario and to Ann Kohen in particular, for bringing together with competence and grace the many fine details of the manuscript and associated correspondence and to Rajni Vaidyaraj for her technical expertise. Tony Moore and Sarah Kramer at Taylor & Francis were a pleasure to deal with and they made the acquisition and production of this book a delightfully straightforward project. The final thanks are due to Catherine Fooks and Eden Anderson, for tolerating excessive work habits and to our children for helping us to understand why we work.

Terrence Sullivan
John Frank
Toronto

Disability at Work: Ministering Solutions for a Wicked Problem

John Frank and Terrence Sullivan

The process of disablement has been a challenging concept for clinicians, worker advocates, industrial sociologists and human rights activists (Jette, 1999). The field is beset with competing models of medical and social definitions of disability (Bickenbach et al., 1999). Most challenging of all, the moral character of disability, including notions of who is worthy and unworthy, has been at the heart of disability discourse from the origins of the disability nomenclature in the evolution of the welfare state (May et al. 1999).

In the specific field of work-related disability, defining disability has been made more complex as most Western nations have witnessed the decline of fatal injuries at work and the cross-over from acute injuries which predominated during earlier industrialization to the longer-latency, insidious-onset, soft-tissue injuries of the musculoskeletal system, associated with the growth of knowledge and service work (Ostry, 2000). These work-related musculoskeletal disorders (WMSD) have less clear causation and often non-specific patho-physiological features. Work-related disability has become more contested as firms struggle to be productive, and they strive to manage down disability-driven payroll costs in an era of global competition driven by trade liberalization (Sullivan and Frank, 2000).

In short, work-related disability is best characterized as a "wicked problem", a notion we have borrowed from Polanyi and Cole (see Chapter 7). "Wicked problems" are those which function in open and interdependent social, environmental and political systems, involving groups with conflicting goals and desires (Mason and Mitroff, 1981). Employers, labor, insurance carriers, health professionals and employees as individuals each characterize work-related disability differently, notwithstanding the fact that most governments must develop rules to grant public disability benefits, with which all these stakeholders must comply. And all of this happens within political systems open to media input and influence.

In our view, for both primary and secondary prevention of WMSD, there has been a substantial failure of the transfer of the best recent research on these injuries to workplace actors, policy makers and clinicians. In our view, this failure occurs for two reasons. First we often search, mistakenly, for the single solution. But research transfer in case of disability must, by dint of the "wickedness" of the problem, be approached simultaneously at multiple levels. Secondly, the transfer fails because, notwithstanding the accelerated transfer, which is arising from the basic sciences in the occupational/public health and health services areas, the research and practitioner communities are relatively inexperienced at transferring practical knowledge to key audiences. Only now are intentional strategies being devised to accelerate this transfer (CIHI, 2001; and see Chapter 9 by Mootz *et al.*).

This book explores barriers and facilitators to the transfer of recent knowledge on disability reduction to workplaces, healthcare providers, insurers and policy makers. Much more can be done to reduce disability at work, if interventions are informed by current evidence, and key strategies to act on this evidence are fully exploited. This book documents the state of evidence as well as creative interventions and knowledge transfer strategies, based on this evidence.

Contrast the challenges in population-level control of two very different health problems: cervical cancer and WMSD disorders. Organized screening of women for cervical cancer may involve one or two challenging issues – such as how to reach high-risk women who do not participate, or how to ensure compliance by screened women in follow up. These are not simple to solve, because they involve motivating human beings, particularly the socially and culturally marginalized (who are most at risk of this disease), to engage in health services which may be unpleasant or frightening. Reducing and preventing WMSD, on the other hand, requires interaction with a complex set of "stakeholders" who typically have – to greater or lesser degrees – competing and conflicting interests.

There is now little disagreement about who needs to be screened for cervical cancer. The challenges are about how to provide the service effectively and efficiently. In the case of WMSD, on the other hand, there is substantial uncertainty and frank disagreement among scientists and stakeholders as to what constitutes "legitimate" WMSD (i.e. disorders *actually caused* by work) and also how best to prevent, treat, and manage these problems.

In spite of the conflicting definitions and various meanings of disability for different stakeholders, as Kerr notes in Chapter 2, recent studies of the etiology of various soft tissue injuries at work have converged on a common understanding. They have strongly suggested independent and primary roles for both biophysical (ergonomic) and psychosocial risk factors in the genesis and prevention of workplace injury. While there are few intervention studies which focus on psychosocial factors in WMSD, there is a large and growing literature on the effectiveness of ergonomic interventions for the *primary prevention* of WMSD. These are highlighted by Norman in Chapter 8 as well as in the highly charged testimony presented in conjunction with the failed US ergonomics regulation (OSHA, 2001; Frank and Lomax, 2002; NAS/IOM, 1999).

Similarly, there is a reasonably good consensus among investigators, based on recent high-quality research, that significant reductions in lost time claim duration *after* injury can be achieved without adverse health consequences for the injured worker (Frank *et al.*, 1998; Sullivan and Frank, 2000). A range of improved disability management practices – secondary prevention – can shorten disability episodes and reduce costs. These new practices are well illustrated in Chapter 4 by

Yassi *et al.*, Chapter 3 by Loisel and Durand, and Chapter 6 by Buchbinder and her Australian colleagues.

In addition to the clinical, workplace and policy studies in this book, others have also shown that modified work can achieve significant reductions in disability as well (Krause *et al.*, 1998). These predictable workplace effects are closely tied to the competitive imperative currently facing nation states and firms, since there does now appear to be a significant effect on aggregate economic output from overall good health (Bloom *et al.*, 2001).

In addition to the arrival of some promising evidence on this "wicked problem", the good news is that, by almost any measure, there have been significant reductions in occupational mortality, and acute work-related injuries, in advanced economies in the past 50 years. The major countries of the industrialized Western world have witnessed declines in serious injuries and fatalities, and the rise of psychological factors in work injury. There is now reasonable evidence that increased psychological distress in the modern world of work is associated with monotonous, low-control work, increased time pressure and pace of work, and a range of adverse health outcomes including cardiovascular problems (Schnall *et al.*, 1999).

Most of these gains in occupational safety and health have been in the upper socioeconomic strata of the workforce (Tuchsen, 1999). There are still significant problems in the least skilled jobs (Sullivan and Adler, 1999). Moreover, there are new areas of concern for injury involving precarious and non-standard forms of employment, forms which are on the rise. Some of these new forms of non-standard work involve "e-lancers" and those who have fared well in the new economy, but many others are those in marginalized low-wage environments engaged in contract work, short tenure employment, multiple jobs, and home work (Quinlan *et al.*, 2001).

So, while overall rates of occupational injury are declining, there remains a disproportionate burden on the disadvantaged and marginalized population. Workers who have not fared well in the "new economy" also include those who have more precarious labor force attachments, those involved in repetitive keyboard work, those working at call centers, and those engaged in lower levels in the financial and banking sector (Payne and Frank, 2000). The listing of priority research areas from the North American Research Agenda (NORA) generated by the US National Institute of Occupational Safety and Health attests to these priority areas (www.cdc.gov/niosh/norhmpg.html).

How then can we deal with these new and wicked problems? This book does not offer the definitive manual for solving the problems of occupational and musculoskeletal injury. Rather, it offers critical insights and practical guidance to those interested in moving forward on the basis of best available evidence, to reduce the burden and impact of these health burdens in the workplace. We suggest that the contributors to this book make the case for the following three general principles to guide understanding and action.

1. WMSD PROBLEMS ARE MULTIFACTORIAL IN THEIR INCIDENCE AND RECOVERY

Not unlike other large public health problems, the forces and risks affecting health and disability at work are inherently multifactorial. The chapters in this book by Kerr and Norman on risk factors, and the chapter by Norman on intervention strategies, identify three broad groups of risk factors: biophysically measured *ergonomic exposures, psychosocial aspects of the work environment*, (usually measured by questionnaire surveys) and the independent factor of *workers' perceptions* of the heaviness of their jobs – "psychophysical" risks.

Similarly, a recent study by Hogg-Johnson *et al.* (1998) and the prognostic literature expertly reviewed by Franche and Krause in this volume, have shown that a range of independent psychosocial factors influence the duration and recovery from soft tissue injury, above and beyond validated measures of "clinical" severity. These include the worker's expectation of recovery, and his or her perceptions of both the workplace response to injury and offers of modified work by the employer. These factors all independently predict the pace of recovery as measured by cessation of compensation benefits. The key point is that workplace disability incidence *and* recovery both involve independent and multiple risk factors, which offer insights as to where interventions might be directed.

Effective interventions to reduce disability at work, as we shall shortly explore, also require multiple levels of "social" interactions between different parties – the worker, the employer, the healthcare provider, and the insurer and policy maker. Any diagnostic approach to resolving worker injury must therefore explicitly involve multifactorial analysis, situating the multiple barriers to recovery and return to work in the social system surrounding the injured worker.

The paradox of disabling workplace injury is that it does not fit the older epidemiological imperative to conceptually separate injury risk analysis from recovery and prognostic modeling. As we noted, most current workers' compensation cases involving timeloss are soft tissue strains and sprains. These conditions typically have subtle onset, often not attributable to a specific traumatic event, sometimes lagged in time of clinical onset, long after initial exposure to ergonomic hazards (e.g. carpal tunnel syndrome in VDT workers). They typically recur over months to years, resolving only slowly and often for no apparent reason. Thus the clean and clear separation of incident, fresh cases and chronic long-standing cases is not possible for many workers with WMSD.

Beyond the blurring of incidence and duration in an epidemiological sense, there is a profound overlap between so-called etiologic and prognostic factors, reflecting the fact that they coexist in the complex social environment around the injured worker. The good news here, of course, is that interventions targeted at such factors can be expected to reduce both incidence and duration of cases – a point made superbly in this volume by Yassi *et al.*

Workers and new forms of work injury do not arise independently in a random fashion, as the "iceberg of disability" image in the Cole and Polanyi chapter so nicely illustrates.

Worker health is distributed and shifts over time dynamically, across a continuum of health-related productivity/wellness at work, which is profoundly influenced by the workplace's entire "culture" concerning health and safety. The

Yassi *et al.* and the Loisel *et al.* chapters provide evidence-based support, derived from workplace studies and experience, for the indivisibility of secondary (treatment of injury to prevent chronicity and recurrence) and primary preventive interventions (reducing the occurrence) in the workplace. Again this is essentially good news: interventions targeted to either outcome are likely to impact on both.

Our current conceptualization of the full set of factors influencing workplace health productivity and disability (Figure 1) depicts a set of forces which conjointly *disable* or *enable* the worker across the spectrum of health-related functioning at work.

Our previous formulations of the work environment and its impact on occupational injury saw a clear separation of the risk factors for injury as opposed to the recovery process, in a linear fashion consistent with epidemiological framing of most health problems (Sullivan and Frank, 2000). In this new formulation, we view the main challenge for interventions to reduce disability at work within a dynamic multi-centred environment, dispersed across the workplace, healthcare system and workers' compensation institutions. Within this dynamic system, an active knowledge transfer process turns on six important categories of interventions. These six intervention categories are listed in the lower right hand corner of Figure 1. This book contains descriptions of examples of all these intervention categories.

These six strategies are given expression through the chapters in this collection. The first strategy – *use of popular interventions*—has long been recognized in adult education and labor circles as being an effective way to transfer knowledge. The chapter by Guzman *et al.* illustrates this technique of small group, multistakeholder workshops as an effective method to convey complex research knowledge. In the case of *provider behavior change* – the second strategy – much of the best evidence suggests that changing provider behavior requires careful systematic approaches focused on influencing the practitioner's social and economic environment (Frank *et al.*, 1998). The *policy reform interventions* are best described by Mendeloff and Thomason in this book, as well as the experimental approaches taken in the state of Washington by Mootz and colleagues. The *disability management approaches* characterized by both Yassi *et al.* and Loisel and Durand highlight the benefit of small incremental approaches which blend workplace and clinical interventions in reducing lost time at work. *Workspace reorganization and design* have always been known risk factors for disability and promising avenues for intervention. The organizational interventions characterized by Norman in manufacturing environments and by Yassi *et al.* in the healthcare sector represent small beginnings for direct workplace action on disabling injury. The review by Kerr and Norman and the intervention chapter by Norman perhaps best characterize the role of ergonomic hazard control, while Mendeloff reviews the limited but compelling evidence on *workplace hazard control* interventions driven by regulatory processes.

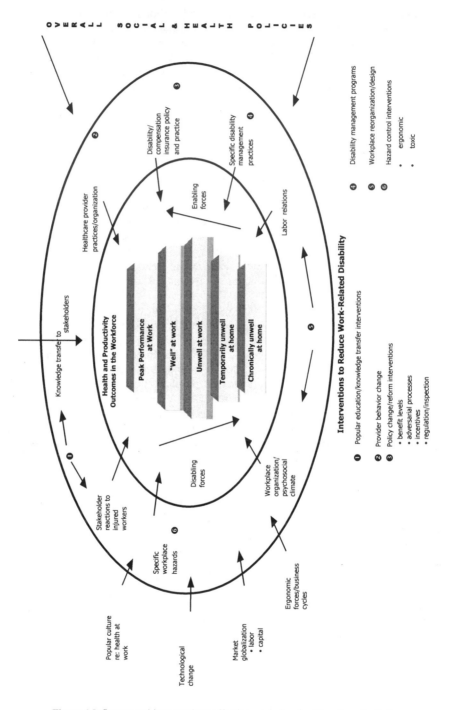

Figure 1 Influence and interventions affecting workplace health and productivity.

2. WORKPLACE SOLUTIONS INVOLVE MOBILIZING STAKEHOLDERS

Will Hutton popularized the idea of "stakeholder society", in which new political institutions are transformed such that the patriarchal rule associated with old style parliamentary structures was being eclipsed in the rule "by stakeholders for stakeholders" (Hutton, 1999). The prevention and management of disability at work is the ideal expression of stakeholder solutions to the wicked problems of disability at work.

The stakeholders are multiple: workers and employers, healthcare providers, public and private insurers (who cover both the rehabilitation services and the income replacement for injured workers), and the policy makers and regulators who set the overall terms of coverage of work injury and the administrative frames for decision making. For workplace action on prevention to be effective, the engagement of virtually all stakeholder groups (workers, managers, healthcare providers, insurers, policy makers, regulators) is critical.

Chapter 5 by Guzman *et al.* illustrates the importance of communication across stakeholder constituencies in conveying an understanding of injury and effective treatment and recovery processes that can lead to collaborative action. Workplace actors, adjudicators and healthcare providers – all must be engaged in a joint learning process. It is not just that they need to assimilate information on how effective and selective workplace factors and treatment processes affect each phase of recovery from injury to return to work. This is important, but only part of the process. Each stakeholder group must also grasp the rational (and irrational) basis for other stakeholders' behavior, so that insightful joint problem-solving can occur.

The very creative media campaign by Buchbinder and her colleagues illustrates the value of educating the public (workers and employers), the providers of clinical care (clinicians and healthcare providers) and the policy makers (workers' compensation boards and insurers) on the scientific evidence surrounding common injuries of the lower back.

Polanyi and Cole by contrast illustrate the important role which action research can play in clarifying conflicts, focusing upon common objectives and engaging stakeholders in a collaborative project to reduce repetitive motion injuries in a large workplace.

Norman, in his chapter on intervention, draws on a rich career of work in manufacturing environments to identify how the risk pathways for major soft tissue injury represent avenues for effective ergonomic intervention. His chapter illustrates that even at the level of the two workplace parties, as a subset of the multiple stakeholders involved in hazard control in the workplace, effective prevention is a challenging task. The challenge is to both understand the competing perspectives of workplace partners and to keep them engaged long enough to evaluate an intervention in a relatively unstable work environment.

The methods employed by Mootz *et al.* in Washington state illustrate perhaps the optimal interactions among workplace stakeholders, occupational health researchers, and state-level workers' compensation officials, for launching and evaluating innovative programs in occupational health services and policies. We suspect no other North American jurisdiction has a fine record of high-quality research, and effective research transfer.

The prospects for incentives and regulation in the policy process, long seen as arch rivals in work injury insurance, are ably reviewed by Thomason and Mendeloff in complementary chapters, illustrating both the limits and the potential of these policy instruments for reducing injury and disability. Although the merits and demerits of these policy tools are hotly contested by key stakeholders, both authors point out converging lines of evidence, from diverse academic disciplines, that suggest effective and efficient policy guidelines are now feasible on the basis of existing studies.

One consequence of a multiple stakeholder orientation is that we should not train people in epidemiology, hygiene, or specific health professional disciplines in this field, without including in their training extensive exposure to the study of social interaction in groups, small group processes, consensus building, collective problem solving via coalition and consensus building, popular education, and the "search processes" required to move forward. To the extent that we view workplace-health trainers' and trainees' skills as largely just technical, we impede complete understanding of how to apply effective interventions. Analogously, public health practice has learned that community development skills are essential for tackling many of the health threats to disadvantaged communities. Occupational health is well on the way to embracing the same principles in its work to reduce occupational disability.

3. DO NOT EXPECT ONE INTERVENTION TO MAKE A DIFFERENCE – SYNERGISTIC AND MULTI-LEVEL INTERVENTION, INVOLVING MULTIPLE STAKEHOLDERS, SHOW THE MOST PROMISE TO REDUCE DISABILITY

The result of much of the work reviewed in this book from Guzman illustrates the value of multi-pronged, multi-stakeholder efforts to reduce disability at work. Polanyi and Cole clearly describe the need to engage stakeholders on issues in causation as well as care and treatment of workplace injury. The efforts described by Yassi *et al.* in Chapter 4, and the work by Norman, as well as the efforts by Mootz *et al.* in the state of Washington, deal with more than narrowly-framed interventions involving a limited focus and range of stakeholders. These interventions inevitably "spill over" to raise broader questions in communities about how work comes to be structured the way it is and how best to reshape it for the benefit of the workforce's health.

The influence of social structures and regulatory frames however can be powerfully influential, since they can drive provider and firm behavior and therefore influence injury frequency and severity, casting a longer shadow which may be superordinate to individual workplaces' particular histories and formative influences.

However, the thoughtful review by Mendeloff underlines an important role for regulations in reducing exposures to toxic substances at work. Even here, his review highlights the fact that many of the US Occupational Safety and Health Agency (OSHA) regulations governing toxic exposures are quite out of date, so that inspection in these particular cases is relatively inefficient, with only 2% of inspections finding overexposures, and 50% of these due to noise, lead and silica.

On the other hand, evidence on the value of regulation and inspection in reducing overall injury rates is less impressive. His review suggests that in big companies, firm-initiated injury reduction efforts have overtaken inspection effects. Alternatively, in dealing with the most egregious toxic exposures, noise and injury situations in the early days of inspection one might inherently expect that detection rates would decline, as inspections go on, making the effects less visible over time. There is a sense here of "vanishing returns" from any particular regulatory policy over time, and a consequent imperative to shift regulatory resources to deal with the emerging threats that each new era brings.

Building on the importance of firm-level initiatives, Mendeloff highlights the value of joint committees with regulatory authority in Canada. His reading of the limited evidence on regulatory efforts to reduce work injury remains somewhat agnostic, with their most demonstrable effects limited to the most egregious injury situations and smaller companies. He does make a forceful case for regulating ergonomic exposures, although there has been little success in introducing such regulation at the national level in the US, due to powerful vested interests and, some would argue, the challenge of devising truly workable legislation applicable to smaller workplaces.

Experience rating to reduce injuries, now the policy instrument of choice, embraced by most jurisdictions, appears to have a small net effect on injury frequency. While surely worthwhile, this is something of a disappointing effect for the scale of effort required to achieve it. By contrast, effective disability management based on evidence may be successful in reducing lost time by up to 50% (Frank *et al.* 1998). Unfortunately, the latter strategy is still encumbered by its image as a "second-best" approach, compared to primary prevention, before injury occurs.

Chapter 9 by Mootz *et al.* describes an effective policy-oriented learning and knowledge transfer environment in the Washington state model, which links researchers and policy makers directly in their efforts to tackle the problems of work-related disability. Critical to this success is the presence of "lynch-pin" personnel who are actively cross-appointed in both government agencies responsible for program and policies and universities that conduct research.

At the regulatory and policy level of intervention, perhaps the good news is that at least some of these institutional arrangements are modifiable, especially when there are mechanisms and methods to achieve stakeholder consensus and then move towards action. The bad news is that when you deliberately try to change large-system organizational features, such as Washington's high-level organizational innovations, only certain outcomes appear amenable to alteration – they could reduce lost time via the use of occupationally-oriented preferred providers (and therefore costs), but not health outcomes or client satisfaction.

The lessons in this volume for reducing workplace disability are highlighted in each chapter. The implications for disability reduction at work are remarkably analogous to the conclusions of an excellent review by Smedley and Syme, identifying the lessons learned from the social and behavioral sciences for improving the health of the population (Smedley and Syme, 2000). Overall, multifactorial, multi-stakeholder and multi-level intervention strategies are necessary to make sustained health improvements occur in a post-industrial population where chronic diseases are predominant.

Analogously, if we seriously wish to take on the challenges posed by WMSD, and improve workplace health and productivity, the lessons are: 1) interventions should focus on the multiple risk factors for WMSD; 2) they must engage key stakeholders; and 3) where possible, they should include multi-level interventions which engage all local workplaces, healthcare systems and societal policy makers in a continuous learning and action cycle.

These may seem ambitious, but in the experience of the investigators and practitioners writing in this book, these are the key evidence-based solutions to address new forms of workplace disability. The new field of workplace disability involves small steps of progress on many fronts to prevent and to manage disability at work. Effective prevention and effective disability management are best achieved not by trying to segregate them in tidy epidemiologic interventions, but perhaps rather by integrating efforts at the workplace.

4. REFERENCES

Bickenbach, J., Challeyi, S., Badley, E. and Ustan, T., 1999, Models of Disablement, Universalism and the Intervention and Classification of Disabilities and Handicaps. *Social Science and Medicine*, **48**, pp. 1173-1187.

Bloom, D., Canning D. and Sevilla, J., 2001, The Effect of Health on Economic Growth: Theory and Evidence. *National Bureau of Economic Research* (NBER), November 2001. NBER Working Paper No. W8587.

Canadian Institute for Health Information (CIHI), 2001, An Environmental Scan of Research Transfer Strategies. *Canadian Population Health Initiative*.

Frank, J., Sinclair S., Hogg-Johnson, S., Beaton, D. and Cole, D. 1998, Preventing Disability from Back Pain. New Evidence Gives New Hope. *Canadian Medical Association Journal*, **158** , pp. 1625-1631.

Frank, J. and Lomax, G., 2002, *Standards of Evidence, the Precautionary Principle in Public Health*. New Solutions, in press.

Hogg-Johnson, S., Cole, D., Frank, J., Sinclair, D.B., Bombardier, C., Brooker, A.S., Clarke, J., Ferrie, S., Haidar, H., Hudak, P., Marx, R., Mondloch, M., Payne, J., Shannon, S. and Smith J. 1998, *Early Prognostic Factors for Duration of Benefits Among Workers with Soft Tissue Injuries*, IWH Working Paper 64.

Hutton, W., 1999, *The Stakeholding Society: Writings on Politics and Economics*. (Oxford: Polity Press).

Jette, A., 1999, Disentangling the Process of Disablement. *Social Science and Medicine*, **48**, pp. 471-472.

Krause, N., Dasinger, L. and Neuhauser, F., 1998, Modified Work and Return to Work: A Review of the Literature. *Journal of Occupational Rehabilitation*, **8**, pp. 113-139

Mason, R. and Mitroff, I., 1981, *Challenging Strategic Planning Assumptions: Theory, Cases and Technique*. (New York: John Wiley and Sons).

May, C., Doyle, H. and Chew-Graham, C., 1999, Medical Knowledge and the Intractable Patient: The Case of Chronic Low-Back Pain. *Social Science and Medicine*, **48**, pp. 523-544.

National Academy of Science/Institute of Medicine, 1999, Musculoskeletal Disorders and the Workplace: Low Back and Upper Extremities.

OSHA, 2001, Testimony by Prof. John Frank, University of California at Berkeley and University of Toronto, regarding Ergonomic Regulation.

Ostry, A., 2000, From Chainsaws to Keyboards; in T. Sullivan (ed.) *Injury and The New World of Work.* University of British Columbia Press, pp. 27-45.

Payne, J. and Frank, J., 2000, Socioeconomic Status and Health-Related Absence from Work; A Canadian Example. *IWH Working Paper.*

Quinlan, M., Mayhew, C. and Bohle, P., 2001, The Global Expansion of Precarious Employment and Consequences for Occupational Health. *International Journal of Health Sciences*, **31**.

Schnall, P., Belbie, K., Landisbergis, P. and Baker, D., 1999, The Workplace and Cardiovascular Disease. *State of the Art Reviews on Occupational Medicine.* (Philadelphia: Hanley and Belfus).

Smedley, B. and Syme, L., (eds) 2000, Promoting Health: Intervention Strategies from Social and Behavioral Research. *Committee on Capitalizing on Social Science and Behavioral Research to Improve the Public's Health*, Division of Health Promotion and Disease Prevention. Institute of Medicine, (Washington: National Academy Press).

Sullivan, T. and Adler, S., 1999, Work Stress and Disability. *International Journal of Law and Psychiatry*, **22**, pp. 417-424.

Sullivan, T. and Frank, J., 2000, Restating Disability or Disabling the State. In T. Sullivan (ed.) *Injury and the New World of Work.* (Vancouver: University of British Columbia Press), pp. 3-24.

Tuschen, F., 1999, Increasing Inequality in Ischaemic Heart Disease. Morbidity among Employed Men in Denmark, 1981-1993. *International Journal of Epidemiology*, **28**, pp. 640-644.

CHAPTER 1

Risk Factors for Musculoskeletal Injury at Work

Mickey Kerr and Robert Norman

1.1 INTRODUCTION

Work-related musculoskeletal disorders (WMSD) are a contentious and challenging research topic, as indicated by the intensity and extent of the debate over precisely what constitutes their main cause (National Research Council, 2001). The debate over WMSD causation can be found within the research community, where the focus often rests with defining the relative importance of different groups of risk factors. Within compensation jurisdictions, policy makers and claims adjudicators, the debate centers around the social policy aspects of the issue, with decisions required to be made despite an uncertain etiologic picture. And lastly, debate is most certainly present within individual workplaces themselves, particularly for workers and managers who must determine the specific set of conditions or events contributing to a given injury. The extent and nature of this multi-faceted debate regarding WMSD etiology is largely driven by two fundamental characteristics of the disorders themselves: 1) they remain the single most important category of work-related health problems in most workers' compensation jurisdictions, in terms of both numbers and costs; and 2) they are very difficult conditions to deal with medically, either to accurately diagnose or to effectively treat, since WMSD sufferers present often without any objective clinical signs.

The challenges faced by the medical profession when trying to accurately diagnose and treat typical, slow onset low back or upper limb pain (Nachemson, 1991), combined with the possible absence of any obvious injury risk factor for the problem, frequently result in skepticism about the legitimacy of many pain reports. Skepticism also surfaces about the relationship of the reported pain to the physical demands of work (Hadler, 1992). This, in turn, leads to speculation and assumptions about what the "real" risk factors are for the reporting of WMSD. Unfounded beliefs about these risk factors are then acted upon in an effort to reduce costs. Some of the more common speculations about the "real" causes of WMSD reports have to do with job dissatisfaction, plant politics, unsafe worker behavior or workers who are too unfit to meet the job requirements. Suspicions about these types of factors being the major contributing causes for WMSD have been around for a long time, no doubt encouraged by a lack of high-quality workplace research on the topic. Widespread acceptance of these worker-based causes for WMSD, in combination with the high costs associated with the disorders, has often led to a reliance upon behavior-based interventions, such as back injury prevention education (e.g. "back schools"), wellness programs or vigorous claims challenging. It is rare to encounter workplace interventions specifically aimed at modifying the work itself, or the way in which it is organized, in order to make the physical or psychological demands of work more manageable. Indeed, comments are more likely to be heard suggesting that "the heavy work has all been engineered out and yet the employees still report injury". In order to inform this widespread debate over WMSD etiology, so that workplace parties considering prevention actions, or decision-makers considering policy changes, can take a more balanced, evidence-based approach to the issue, this chapter will present a review of the workplace research base concerning WMSD etiology. There are many things that might be contributing to the onset of WMSD disability as summarized in Figure 1.1.

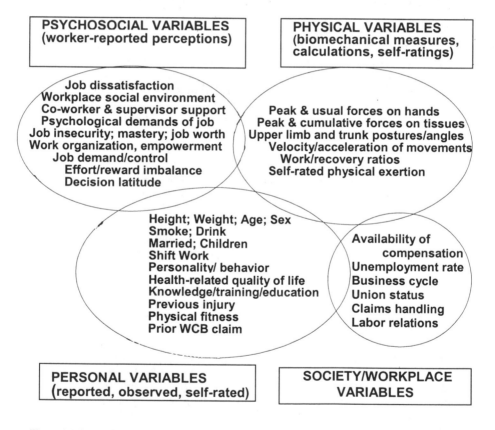

Figure 1.1 Categories of hypothesized risk factors for reporting WMSD. While some factors shown are proven risk factors, many are not. Society/Workplace factors are not directly addressed in this chapter.

This is not an evidence-based picture of WMSD etiology however, as it reflects a combination of hypotheses, preconceptions and everyday biases, as well as research evidence. By the end of this chapter, we hope to have left readers with a better understanding of which of the factors shown in Figure 1.1 are indeed contributing to WMSD disability. This chapter should therefore provide the reader with a solid foundation that should, in combination with the rest of the chapters in this book, and in particular with Chapter 8 on interventions by Norman, leave the reader well positioned to effectively contribute to efforts aimed at reducing the extensive burden of WMSD.

1.2 BACKGROUND

Despite the recent overall reductions in disability claims rates seen in most compensation jurisdictions, soft-tissue sprain and strain-type injuries still typically account for well over half of all compensation claims and an even larger proportion of the direct costs of providing workers' compensation disability benefits. Recent events at the policy level in the United States, where regulations aimed at preventing WMSD were first enacted, and then immediately repealed by outgoing and incoming administrations respectively, clearly underscore the fact that this causation debate is far from resolved (CTD News, 2001).

Certain aspects of the WMSD issue, especially in relation to the filing of compensation claims, lie in the realm of social and economic policy and are thus outside the scope of this chapter. We will restrict our discussion here to one of disability causation, where much of the concern has focussed on the relative contribution of biomechanical and psychosocial factors to the onset of WMSD pain. Some researchers have argued that one or the other set of risk factors is the predominant cause of WMSD, leaving little room for compromise (Frank *et al.*, 1995). Yet several studies have now shown independent effects for these as well as other types of risk factors (e.g. individual characteristics). In view of the debate surrounding the work-related etiology of WMSD, a summary of the recent published literature on the topic is presented to assess the extent to which different types of factors may be contributing to the onset of WMSD. We also present one study in some detail as it nicely illustrates both the complexity of WMSD etiology and the considerable challenges faced when attempting to conduct high-quality workplace research. While the focus of the chapter is the role of work-related factors, especially the relative contribution of biomechanical versus psychosocial factors, it should also be recognized that there are several other factors that have been presumed to influence the onset of WMSD, including individual characteristics such as age or gender, and societal characteristics such as unemployment rates or the extent and availability of disability benefits. Given the overall focus of the book however, this chapter focuses deliberately on the types of exposures deemed most relevant to occupational settings. Since back pain and upper extremity disorders account for the vast majority of WMSD in virtually all workers' compensation jurisdictions, this review focuses primarily on the risk factors for these two body regions, with additional emphasis given to discussing back pain in particular, given the particularly high prevalence of this problem.

The chapter is also largely restricted to reviewing the results of epidemiologic studies rather than laboratory studies, as the former are often presumed to provide the strongest level of evidence needed to inform the current debate on causation. While a more comprehensive review of this topic would cover laboratory-based biological pathway literature in order to properly interpret the correlations observed in epidemiological studies, such a comprehensive review would require an entire book itself, and not just a single chapter. While well suited for uncovering potential WMSD risk factors, workplace epidemiologic studies are not well suited to specifying processes of causation. In vitro loading studies of spinal motion units and, more recently, cellular level studies of the effects of repetitive low-level loading on muscles, are more properly suited to identifying causal mechanisms of tissue damage resulting from physical loading. In addition, biomarker studies, such as salivary cortisol responses to laboratory-controlled stress tests, are also needed to better understand possible causal mechanisms associated with workplace stress.

Unlike some exploratory epidemiologic studies, the lab-based studies are not correlational studies, but rather tissue damage or biological causation studies that refine and test specific hypotheses related to cause and effect within a controlled experimental model paradigm. Understanding the causal models developed and tested in the laboratory and then transferring them into "real world" evidence, in the form of rigorous workplace epidemiologic risk factor studies, remains an obligation for occupational health researchers interested in contributing to either policy making or evidence-based prevention efforts. However, primarily for cost reasons, many workplace studies, in particular the highly regarded prospective epidemiologic studies, often end up relying upon self-reports, job titles or other relatively error-prone sources of exposure data in order to economically gather data on the large numbers of subjects typically required in such studies. The resulting error associated with measuring occupational exposures in these studies creates a substantial problem, even for relatively easily-diagnosed conditions such as cancer. However, it creates a huge problem for conditions such as low back pain (LBP) and upper limb pain that are not as reliably diagnosed by physicians.

Despite these measurement challenges, many policy makers and claims adjudicators often demand epidemiologic data and disregard laboratory data in their decision making. A broadening of what constitutes "evidence" may be required when dealing with complex conditions like WMSD. It is incumbent upon epidemiologists, policy makers and claims adjudicators to understand both the biological pathways and epidemiologic literature so that decisions and possible interventions and risk-factor studies will be well informed by having a solid conceptual foundation. Similarly, laboratory-based researchers need to recognize the complex etiologic pathways present in these conditions in order to work with epidemiologists to develop and refine feasible workplace exposure assessments that are based on the causal pathways they have developed and been tested in the lab. While it can create additional challenges, such broad-based interdisciplinary work will help everyone to better understand the significance of any potential risk factor relationships uncovered in workplace studies. While acknowledging the importance of such a cohesive synthesis of the available etiologic evidence, limitations in the scope of this chapter leave our review focussed on the epidemiology literature, as notwithstanding the tensions outlined above, this literature probably remains the most persuasive influence affecting compensation policy. However, it is important that the reader also becomes familiar with the biological pathways literature to get a more rounded picture of WMSD causation (Kumar, 2001; McGill, 1997; Adams and Dolan, 1995).

This chapter draws heavily upon several recent reviews on WMSD etiology, but, it should not in itself be considered a structured, critical review of evidence concerning the work-relatedness of these disorders, as has been recently published elsewhere (Bernard *et al.*, 1997; National Research Council, 2001). The reader is directed to more extensive reviews on psychosocial risk factors for WMSD that have been published recently, as have additional articles reviewing the epidemiology of WMSD, including some specific to LBP and upper extremity disorders (Bongers *et al.*, 1993; Kerr, 2000; Hoogendooorn *et al.*, 1999; Hoogendooorn *et al.*, 2000; Malchaire *et al.*, 2001). We hope to provide the reader with a general understanding of current knowledge on risk factors for WMSD, particularly as they relate to the issues raised in the following chapters of this book. The information contained in this chapter on the causation of WMSD is of special relevance to the discussion of workplace interventions for the prevention of injuries found in Chapter 9. Some researchers have argued that the ubiquitous nature of musculoskeletal disorders implies that intervention efforts aimed at primary prevention of WMSD are likely to be futile and, hence, by extension, attempts at understanding their etiology are largely irrelevant as well. This book will present a different argument, one that supports workplace prevention efforts through a better understanding of what causes WMSD. Given the fact that most people who develop WMSD will also likely experience a recurrence of the condition at some point in time, there is some element of futility in debating primary versus secondary prevention. It is very likely that those factors that contribute to the onset of the condition also reduce the likelihood of returning to work, and vice versa. In fact, workplace interventions that are aimed at reducing the extent or likelihood of exposure to risk factors for LBP and upper limb disorders (ULD) should have the effect of reducing disability, facilitating return to work (RTW) and reducing primary risk all at the same time. How each study selects and defines its WMSD outcomes is therefore an important consideration, as the more specific and less common the condition under study becomes, the less applicable the results are to typical, non-specific workplace WMSD. The congruence between casual pathway results and epidemiologic results can collapse over this point, as what causes the initial tissue damage may not be the same as what causes re-injury, especially in relation to absolute exposure levels. Further details on how to best interpret epidemiologic studies on the etiology of WMSD can be found elsewhere (Bombardier *et al.*, 1994).

Lastly, because the chapter is structured along exposure categories rather than specific musculoskeletal conditions, it implies a certain degree of overlap in the risk factor categories across the spectrum of conditions comprising WMSD. This does not, however, mean that specific sets of cause-effect relationships do not exist. To the contrary, such specific relationships clearly do exist. For example, spinal disc compression is a relatively well-known risk factor for LBP, but it is clearly not part of the causal pathway for a wrist problem like carpal tunnel syndrome. While identifying these direct exposure-disease relationships may be helpful in specific diagnostic or workplace situations, the infinite number of unique workplace settings and conditions argues for a more general approach that informs readers at a broader level. Such is the approach adopted in this chapter.

1.3 RISK FACTORS

Winkel and Mathiasson (1994) describe three categories of risk factors for WMSD, such as back pain: 1) individual factors, such as a person's weight, a history of prior LBP, and smoking habits; 2) physical factors, such as lifting and posture (referred to in this chapter as biomechanical risk factors); and 3) psychosocial factors, such as job control and job satisfaction. Given recent findings in some studies, including a recent study of back pain in automobile assembly workers, it may be worthwhile now to add another category, somewhere between the physical and psychosocial groupings (Kerr *et al.*, 2001). The additional category would include the way in which workers perceive the workplace demands they experience while at the job, over and above how they may be measured objectively. These kinds of measures are referred to as psychophysical factors, and include workers' self-ratings on how physically demanding the job is on the body. Given these measures' more explicit connection to the physical demands of work we have included them here under the physical category for the sake of simplicity. The epidemiologic evidence concerning the more frequently encountered examples of each of the three main risk factor categories is briefly summarized below. The relative strength of the evidence supporting the more established of these risk factors' relationships with WMSD is also outlined in Table 1.1.

Table 1.1 Independent risk factors for the occurrence of WMSD
(modified after Bernard, 1997; and Frank *et al.*, 1995).

Strength of evidence[*]	Individual risk factors	Physical and biomechanical risk factors	Psychosocial risk factors
Weak	Smoking Obesity (or high body mass index) Female Strength	Low static load	Job dissatisfaction Low social support Monotonous work Effort-reward imbalance
Moderate	Age	Heavy self-rated physical exertion Postural stress (e.g. overhead work or awkward trunk angle) Repetitive motion Cumulative load	High (psychological) job demands Low job control
Strong	Prior medical history (injury)	Forceful exertions Heavy lifting Vibration	N/A

[*] "Weak" evidence indicates that published evidence has been inconsistent and thus inconclusive, but suggestive of some increased likelihood of developing WMSD if exposed. A "moderate" level indicates that some convincing published evidence is available for an association with the exposure. A "strong" level indicates that exposure to the factor is very likely to increase the risk of developing WMSD, based on consistency and strength of the research findings. N/A indicates that no risk factors are found within the category. Entries in the table are considered "independent risk factors" as the evidence to support their inclusion is generally based on multi-variable analyses.

1.3.1   Individual Factors

Prior Medical History (or Previous Injury)

The factor with the most consistent association, and perhaps the only well "established" WMSD risk factor, is a history of prior pain (or injury). For example, previous episodes of LBP have been shown to be predictive of who will report it in the future (Battie and Bigos, 1991). Due to the lack of any clear underlying tissue pathology for most WMSD, including LBP, the mechanism for this association remains unclear (Adams and Dolan, 1995). As a direct biological influence, it may be indicative of a reduced threshold for injury, or pain, for tissue that has been previously damaged. As an indirect marker of exposure, the association between prior injury and future injury could also be the result of a high-risk exposure history that has remained consistent over time. For example, if a physical workload was contributing to the onset of LBP, and it remained unchanged at the LBP-inducing level, then LBP risk would obviously remain at an elevated level as well, resulting in repeated LBP episodes. This hypothesis remains the subject of conjecture however, since no studies have been able to document long-term cumulative exposure history accurately enough to control for it in an epidemiological analysis, nor is the appropriate exposure metric yet known to permit the execution of such a study (Burdorf, 1992).

Age

The association with age is not well understood, but the results of studies have shown some consistency for the relationship in younger and middle-aged adults, with risk appearing to increase with age in this range (Liira *et al.*, 1996; Houtman *et al.*, 1994). However, this association often disappears when studies control for other factors, such as physical workload (Burdorf and Sorock, 1997). The weakness of the possible association between age and WMSD is also demonstrated by the fact that some studies have reported a negative association, particularly among older workforces, where the younger adults were more likely to file a compensation claim or report the problem at work (Abenhiem and Suissa, 1987; Zwerling *et al.*, 1993). It is also common for the incidence of WMSD compensation claims to be higher in younger than older adults, especially for males. This apparently increased risk of developing WMSD in younger workers may be an effect of heavier workload or lack of experience, though neither of these possible explanations was fully accounted for in the observed studies. In any event, there is no clear relationship between age and risk of developing WMSD. What confounds the issue even more is the changing nature of work as one gets older, especially in seniority-dominated work environments, where older workers tend to migrate towards lighter work. This simple look at the age incidence of WMSD within such environments will be clearly confounded by the uneven distribution of workplace exposures. It is also possible that older workers with a strong history of WMSD may eventually leave the workplace altogether on account of their pain, especially if gradual accommodation to lighter work with age is not possible. This attrition of WMSD-afflicted workers could leave the "survivors" being unrepresentative of workers at large ("healthy worker effect"), which could again bias or confound conclusions about WMSD risk for a given workplace (Bernard 1997).

Other Factors

Even less well-established risk factors, such as, obesity, cigarette smoking and female gender, have all been suggested as possibly increasing the likelihood of developing WMSD. The strongest evidence in support of an association between smoking and WMSD can be found in the LBP literature, where several studies have noted this relationship, although the strength of the association is often weak (i.e. the amount of increased risk reported was small) and possibly confounded (Goldberg *et al.*, 2000). For both female gender and obesity (or its related measure a high body mass index), there is equivocal evidence both for and against a causal relationship, especially in relation to upper extremity problems such as carpal tunnel syndrome (Bernard, 1997; Power *et al.*, 2001; Malchaire *et al.*, 2001). Finally, there is at best contradictory epidemiological evidence supporting an association between the occurrence of WMSD and general fitness, physical activity or muscular strength level (Bernard, 1997; Hoogendoorn *et al.*, 1999; Malchaire *et al.*, 2001).

   Despite this limited evidence in support of individual characteristics being implicated in the causal pathway for WMSD, much of the effort in preventive measures has been focused here. As discussed in more detail later in Chapter 8 by Norman, only a few of these suspected individual risk factors are readily modifiable, especially in the workplace, and thus are of limited potential utility in prevention strategies specific to WMSD, although fitness and anti-smoking campaigns certainly have considerable value in the prevention of other health outcomes, most notably cardiovascular disease and cancer. As a final note, abnormalities observed on radiographs, such as those used for pre-placement back pain screening programs in workplace settings, have now been clearly shown to be of no or limited value in predicting who will develop WMSD (Gibson, 1988; Cohen *et al.*, 1994). It appears as though attempts at identifying spinal deformities associated with typical LBP, even when using the most modern and sophisticated imaging equipment, are of no more value in determining etiology than they are of diagnostic value (van Tulder *et al.*, 1997).

1.3.2 Biomechanical (Physical) Factors

Assessing Exposure

Most of the published literature examining biomechanical risk factors for WMSD consists of cross-sectional studies, such as surveys taken only at one point in time (Bombardier *et al.*, 1994). The use of error-prone measurement instruments, such as indirect (subjective) exposure measures from self-reported questionnaires, and overly simple exposure categories based on job titles, are common features of WMSD etiologic research (Winkel and Mathiassen, 1994). For several reasons, but probably most likely due to their relatively high cost, studies using direct (objective) measurements of biomechanical exposures in the workplace are the exception rather than the rule (Burdorf, 1992; Garg and Moore, 1992; Hoogendoorn *et al.*, 1999). An example of one recent study that specifically addressed this measurement error concern is described in some detail below, in order to better examine the relative contribution of physical/biomechanical and other workplace risk factors for LBP (Kerr *et al.*, 2001).

The Complexity of WMSD Etiology – a Recent Example from the Back Pain Literature

As is the case for many WMSDs, the causes of LBP are not well understood. Typically, there is no clear evidence to medically pinpoint the underlying problem in the vast majority of cases. The exact nature of the tissue damage in many common musculoskeletal conditions, such as LBP, often remains undetectable even by the most advanced medical imaging methods, and physical examination is often uninformative. Health professionals, as a result, must rely on the patient's description of symptoms as the chief means to diagnose the condition. This is a situation that both physicians and patients find frustrating. Add to this the work-related disability that often accompanies LBP, and a forbidding environment of suspicion and mistrust can be created – regarding not only the presence of the pain itself, but also whether or not the pain is a consequence of the work being done.

The Ontario Universities Back Pain Study (OUBPS), therefore, set out to determine if workers in physically demanding jobs (i.e. exposed to biomechanical risk factors) were really at greater risk of developing LBP; or if LBP was more likely a consequence of a work environment that was perceived by the worker as being unsupportive (i.e. there was exposure to psychosocial risk factors); or if both sorts of risk factors contributed to the problem. At the same time, individual characteristics were measured so as to control for them in the analysis, so that the focus could remain on determining the contribution of various work-related factors to the onset of LBP. Previous research had been of limited use in answering these questions since it focused almost exclusively on only either the biomechanical aspects *or* the psychosocial aspects of work related to LBP, but rarely measured.

How Was the Study Done?

To collect sufficient data in a reasonable time period for this study, a large population of workers at significant risk of reporting LBP was required. The study was conducted at a large automobile assembly facility that employed approximately 10,000 hourly-paid production and support workers in the manufacturing of cars and light trucks. "Cases" of newly reported LBP were collected from the occupational health nursing stations spread throughout the plants. The "controls" were randomly chosen from a roster of the entire current workforce, as the cases presented. A "case-control" study design, which enrols workers *after* they get LBP and compares them to unaffected workers, was used instead of the "cohort" study model, which would examine people *before* they got

LBP. While potentially more prone to bias than a cohort study, the case-control approach was adopted for several reasons. The "cohort" design can be very expensive when many risk factors need to be monitored and examined over time. Thousands of healthy workers would have to be included in a cohort study in order to ensure that a sufficient number of people with LBP would eventually be part of the study. This would have severely restricted the amount and type of information that could be collected, since repeated, detailed measurements of the physical demands of the job, for example, would be impossible on so many study subjects because of the high cost. If only baseline measurements were used, exposure information collected at the start of the study would likely be well out of date by the time the required years of follow-up time had passed. This potential for misinformation could severely diminish the advantage usually conferred by a prospective cohort study design. A "case-control" study, with detailed and accurate assessments of the workplace risk factors was thus a viable choice. With a focus on very recent LBP reports to the nursing stations, use of standardized questionnaires for psychosocial risk factor assessments, and the comprehensive and objective nature of the assessments of the physical demands of the job – all helped to ensure that collecting data after LBP had occurred did not bias the study results. When carefully designed and executed, case-control studies can provide evidence that is as good as those provided by cohort studies (Rothman, 1999). Figure 1.2 provides an outline of how the Canadian study was conducted.

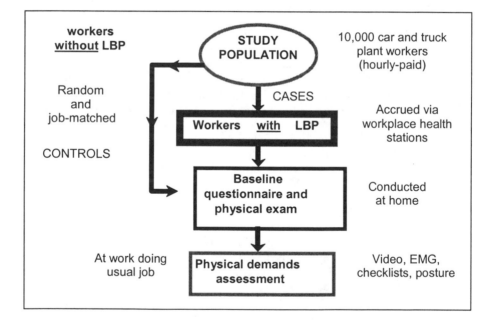

Figure 1.2 Outline of the Ontario Universities Back Pain Study (OUBPS).

Who Was Studied?

"Cases" comprised those workers who experienced a recent LBP episode that was reported to one of the onsite occupational health nursing stations. Cases were screened to ensure that they had not reported LBP in the last 90 days and that they did not have any other serious medical problems. To be part of the study, cases did not have to file a workers' compensation claim nor did they have to have missed any workdays because of LBP. When contacted by the study's investigators, approximately 70% of workers reporting LBP agreed to participate in the study, with 137 eventually being enrolled as eligible cases. As cases accrued through the nursing stations, the main "control" group was assembled through a random selection of healthy workers who had not reported LBP to the worksite in the past three months. About half of the potential controls contacted agreed to participate in the study, with 179 workers eventually being enrolled as participants. A second smaller group of controls was matched to cases as exactly as possible, based on the job the case had at the time of reporting LBP. Based on direct observation by trained observers, the matching focused closely on the physical tasks and output of the work being done. In total, 65 "job-matched controls" were successfully recruited for the study. The main purpose of this group was to provide biomechanical data for cases who for logistic reasons could not have the physical demands of their job assessed, and to permit a comparison of questionnaire responses between injured and uninjured people doing the same job, to insight into how the injury process itself might have affected responses to certain questions.

What Was Studied?

Data on a wide variety of basic demographic and health-related characteristics of individual workers, as well as their work activities and their perceptions of workplace environment, were collected for the study. The information presented in this report was derived from detailed biomechanical assessments of subjects performing their normal job at the worksite, as well as an interview-assisted questionnaire and a brief physical exam, both usually completed in the worker's home.

Physical Demands Assessments

The assessments of the physical demands of work focused on quantifying the biomechanical forces present in the lumbar spine when workers performed their usual tasks. Measuring or calculating the forces produced – rather than describing the tasks being done – was carried out because it would help to ensure that the results of the study could be applied, in the future, to the widest possible range of workers, and not just those at General Motors or other automobile plants.

The biomechanical assessments "on the job" included measures of estimated lumbar disc compression, shear forces acting perpendicular to the spinal column, the largest load the worker handled during the assessment, and several aspects of back posture and trunk motion. The list of variables examined for the main risk-factor analysis was only a sub-sample of a much larger "physical demands data set", collected in order to help further develop field methods to be used in future workplace epidemiological studies.

Self-reported Data

In addition to the physical demands assessments, participants received a comprehensive interview-assisted questionnaire that addressed demographic and psychosocial factors. Clinical data on back pain and other health problems were also collected, and all subjects were given a brief clinical exam to assess LBP-related symptoms. The study asked workers about the amount of control they perceived having in their jobs, as well as their perceptions about the psychological demands of their job. Another important factor examined was the degree of social support workers felt at work, especially from supervisors and co-workers. Based on previous studies, these three factors – control, psychological demand and social support – were considered to be of particular importance. The study also examined perceptions of job satisfaction, highest education level achieved compared to others in a similar job, how valued the worker felt, and levels of empowerment and personal control. Workers were also asked to rate how physically demanding they perceived their jobs to be.

How Were the Data Analyzed?

"Cases" and "controls" were contrasted on each of the main study variables, or potential risk factors, separately at first, and then in combination. Multi-variable logistic regression was the main statistical tool, since it allows for identification of the independent effects of each of the possible risk factors in a given model, while controlling simultaneously for the effects of the other variables being examined.

OUBPS Main Findings

Figure 1.3 is a visual summary of the main results from the OUBPS. The results of this study demonstrate the importance of considering a wide variety of factors, as evidenced by the fact the all three types of factors were identified as being independent risk factors for LBP. By examining all three categories of factors simultaneously in the same statistical model we were also able to obtain a visual impression of the relative contribution to the risk of reporting LBP for each of the risk factor categories, underlying the complexity of their possible inter-relationships. The results of the combined multi-variable analysis confirm that biomechanical and psychosocial risk factors both have an independent role to play in the reporting of LBP at the study site.

Based on the final statistical model developed by the study, the most important biomechanical risk factors for LBP are peak shear force, peak hand force and the amount of disc compression accumulated over a full shift. The influence of the other biomechanical risk factors disappeared when subjects' combined risk factor profiles were controlled for, because many biomechanical exposures were correlated with each other, measuring very similar concepts. This is a practical benefit, in that interchanging some, biomechanical variables for others, perhaps more easily measured, does not lower the model's predictive ability appreciably.

Odds ratios

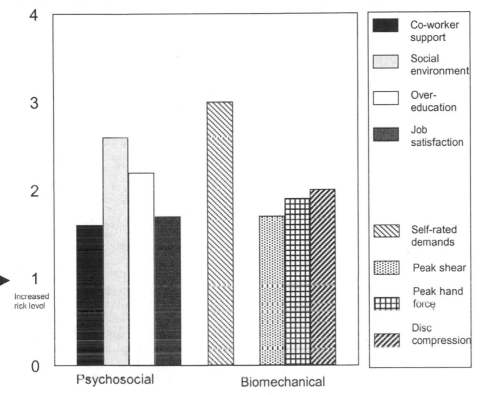

Figure 1.3 Statistically independent risk factors for the reporting of LBP in a large auto assembly plant (Kerr *et al.*, 2001). The individual factors adjusted for included age, smoking, marital status, pre-school children, body mass index and prior compensation for LBP. Only the latter two were significant risk factors, and both roughly doubled the risk for reporting LBP.

According to the final model, workers with high levels for *any* of the three statistically significant biomechanical variables had an increased risk of LBP that amounted to almost a doubling of the risk for workers without the exposure. However, if a subject's exposure was high on all three factors, the LBP risk increased approximately five-fold, regardless of the presence of any other risk factors. This risk, associated with the physical demands of work, was further increased if one included the self-rated measure of physical exertion used in the study, which showed an even stronger independent association with LBP status.

The most important psychosocial factors appear to be the workplace social environment and the self-assessed relative fit between the job and the education of the worker. The size of the risk effect for these factors varied somewhat, with the strongest ones, such as a poor workplace social environment, more than doubling risk. It is interesting to note that there appears to be a fairly strong effect for job control only after the physical demands of work are taken into account. This effect did not appear at all when "job control" for cases and controls was examined by itself. This is a strong indication of the complex inter-relationship that exists between the biomechanical and psychosocial work environments. In contrast with some other studies, workers who were more satisfied and reported better co-worker support were somewhat more likely to report LBP. Since cases were a mixture of workers with and without a work absence due to LBP, the unexpected direction of these results could be an indication that people who like their work will remain at it, despite their pain, and that the support of their co-workers may be helping them to do this. As a final point, the consistency of the findings for the physical load measures was striking, as every single variable examined showed cases with higher exposure levels than controls. The same consistency of effects was not seen for the psychosocial factors, as indicated above, suggesting that their effects may vary from setting to setting.

At each step in the final analysis, the appropriateness of the statistical tests used was carefully examined. The exclusion of 20 cases in which the biomechanical data from a job-matched control had to be used did not appreciably change the results shown here. The results were also largely unaffected by excluding the 61 random controls whose questionnaire responses suggested that they might have had LBP but had not reported it in the last three months. The results of these sub-analyses, and the monitoring of the multi-variable statistical procedures used, suggest that the final statistical model does a good job of accounting for the risk factor differences observed between cases and controls, explaining about 44% of these differences, and correctly "predicting" subjects as cases or controls about 80% of the time. Both of these indices of model accuracy are quite high values for an epidemiological study of this kind. In other words, the final model can legitimately be used as the *basis* for prevention efforts since it seems to "explain" much of what distinguishes workers who report LBP from those who do not.

Significance of the OUBPS Results

This study confirms the multi-factorial origins of occupational LBP, and thereby provides some guidance for the development of workplace interventions aimed at preventing the condition. Because the study chose specifically to use directly measured biomechanical components of work, rather than simply relying on self-reported loads, or rogue group-level measures such as job title, it provides better estimates than most studies of the influence that the physical demands of work have on the occurrence of LBP at work. These risk estimates have also been "adjusted" for the independent effect that was observed for some of the psychosocial risk factors, using data collected from a structured, interview-assisted questionnaire. Despite some international expert opinions suggesting that interventions should be aimed only at preventing disability after LBP has already been reported, and that these interventions should focus on the worker themselves rather than the workplace, since musculoskeletal pain is an "inevitable part of working life", the identification of potentially remediable risk factors in this study suggests that primary prevention, before the onset of disability, is indeed feasible.

Specifically, this study has identified several important biomechanical and psychosocial risk factors related to the reporting of LBP in a large manufacturing facility. It has resulted in the development, validation and use of several biomechanical "tools" on the plant floor, used to measure physical demands of more than 300 jobs. Some of these tools have been transformed from research methods to practical instruments for more general use in the workplace. In addition, work is in progress to attempt to identify particular levels of exposure to risk that could be used to establish job design "target values".

The next step is to use the information obtained from this and other new high-quality research studies to develop practical interventions that would address characteristics of both the psychosocial *and* physical work environments, so as to reduce risk. This study suggests that efforts to address only one class of risk factors will yield less impressive results for the prevention of LBP at work than efforts aimed at a wide range of working conditions.

We now know the relative contribution of several risk factors to the reporting of pain at work and how to reliably measure all of the variables. Unfortunately, this study has also shown that there is no single "magic bullet" risk factor that, if removed, will significantly reduce work-related disability. No single variable, by itself, accounted for a great deal of the pain reporting but all of the variables combined accounted for nearly 45% of the pain reporting. This knowledge is an appreciable advantage in designing an effective intervention program. However, as discussed later in Chapter 8 by Norman, it also shows that no single intervention such as a wellness program, a more supportive workplace environment, or a better tool design, is likely to make much difference. This is not unlike the situation faced by public health efforts to reduce other multi-factorial diseases, such as heart disease and cancer.

1.3.3 Psychosocial Factors

There is limited but growing empirical evidence linking psychosocial risk factors to the occurrence of LBP. Despite concern that the role for psychosocial factors may be only indirect, through increased exposure to the physical demands of work, rather than a direct causal link *per se*, evidence in support of a direct role is beginning to emerge. A recent review identified a number of variables, including monotonous work, high perceived workload and time pressure, as being possible risk factors for musculoskeletal disorders, including LBP (Kerr *et al.*, 2001). The possibility that psychosocial factors may have an independent contribution to the etiology of WMSD has been substantiated by several studies, including an analysis of routinely collected survey data in the Netherlands, reporting an association between psychosocial stressors (high "work pace" and lack of intellectual discretion) and musculoskeletal complaints including back pain, even after taking (self-reported) physical work stressors and worker characteristics into account

(Houtman *et al.*, 1994). Physical workload demonstrated a strong independent association with LBP on its own as well, in addition to the aforementioned work pace, which could itself be a marker of physical as well as psychological load. More recently, a prospective cohort study of 721 workers at 226 automobile repair garages by Torp *et al.* (2001) has also concluded that psychosocial factors at work, such as low job control, may predict musculoskeletal pain.

Further support for a link between psychosocial factors and soft-tissue musculoskeletal problems at work comes from a study by Faucett and Rempel in video display terminal (VDT) operators at a large US newspaper (Faucett and Rempel, 1994). Like the Dutch study described above, these authors reported separate, independent associations between biomechanical aspects of work (keyboard height and posture), job-related psychosocial factors (workload, decision latitude and support) and the presence of musculoskeletal symptoms. However, as with the Dutch study, the strength of the association was generally higher for biomechanical factors than that observed for psychosocial factors. This research has highlighted the importance of examining the interplay between psychosocial and biomechanical risk factors.

Finally, the Boeing Study, possibly the most detailed LBP research effort yet completed, is often cited to support the association between job-related psychosocial factors and LBP to the exclusion of biomechanical variables as risk factors (Bigos *et al.*, 1992). The main finding of this prospective three-year cohort study was that, apart from a prior history of LBP reports, worker dissatisfaction with the job tasks was the only work-related risk factor associated with subsequent reporting of LBP. None of the variables assessing the physical demands of work significantly predicted who reported back pain. However, it is possible that the ability of this study to identify biomechanical risk factors was limited by a misclassification error resulting from the assignment of exposure data from *group-level* biomechanical assessments to the study subjects, rather than the collection of data for each individual doing his/her own job.[1]

It should also be noted that the final statistical model in this study could predict only a small proportion of the observed cases, even when the effects of all measured risk factors were combined. Given these caveats, and the paucity of corroborative studies, it may be premature to conclude that job dissatisfaction *per se* is a more important predictor of LBP than the biomechanical demands of work (Volinn *et al.*, 2001).

1.3.4 Risk Factor Summary

This chapter has focussed on summarizing the strength of the evidence concerning the more commonly encountered risk factors for WMSD – i.e. the ones that appear to have the most consistent published evidence supporting a link between exposure and the onset of the condition. While many possible risk factors were discussed, it should be noted that very few of these would be considered as "well-established" risk factors, based on the standard epidemiologic criteria of causation (Bombardier *et al.*, 1994). Although the overall quality of the published evidence on risk factors can perhaps be considered limited, since much of it is cross-sectional in nature, it seems fair to say that there is now sufficient evidence to conclude that workplace exposures are contributing to the onset of WMSD. This conclusion is especially appropriate in relation to the more recent studies examining the issue of measured biomechanical load as a possible WMSD risk factor.

The scientific rigour of these recent studies is substantially greater than that of their earlier counterparts, given the more sophisticated study designs and/or the extent and quality of their exposure assessments. Although there has been considerable development in the research methods, and in the elucidation of some of the causal pathways for certain conditions, many questions remain about WMSD etiology. Hence there remains a need to identify specific, modifiable components of work that are causally associated with the problem, so that effective interventions can be mounted. Future research should therefore focus on making detailed measurements of workplace exposures, from both the biomechanical and psychosocial realms, so that the effects of these components can be disentangled. The latter category of risk factor is of particular importance as a research topic, given the mounting evidence for a role for psychosocial factors in WMSD etiology (Kerr *et al.*, 2001). Recent laboratory work by Marras *et al.* (2000) that examined the biomechanical consequences of increased work stress on the lumbar spine is a good example of the kind of work needed to help bridge the gap between basic, laboratory research on WMSD causal pathways, and the risk factors identified observed in workplace epidemiologic studies. A better understanding of the causal significance of these factors, and the potential inter-relationships between them and the physical and individual risk factors, will undoubtedly help us develop effective intervention strategies that take a comprehensive approach to the problem.

1.4 KEY MESSAGES

- Work-related musculoskeletal disorders (WMSD) are the single most important category of work-related health problems in most workers' compensation jurisdictions, in terms of both numbers and costs.

- WMSD are very difficult conditions to deal with medically, either to accurately diagnose them or to effectively treat them, since WMSD sufferers present often without any objective clinical signs.

- The misconceptions generated by the first two points above have led to considerable debate and uncertainty regarding the etiology of the more common WMSD.

- While it can create additional challenges, broad-based interdisciplinary work to unravel the complex findings uncovered in workplace studies is necessary to help everyone better understand their potential significance.

- The strength and consistency of the findings from studies linking workplace biomechanical factors with WMSD is convincing evidence for the work-relatedness of these conditions.

> • Work-related biomechanical and psychosocial factors, as well as certain characteristics of individuals, particularly prior episodes of pain, have all been shown to have an independent role to play in the reporting of WMSD.

1.5 NOTES

[1]Biomechanical data were collected for the Boeing study only for "job types" employing 20 or more workers. The remaining subjects were assigned the exposure values based on this site-specific job classification scheme (see Bigos *et al.*, 1992).

1.6 REFERENCES

Abenhaim, L. and Suissa, S., 1987, Importance and economic burden of occupational back pain: a study of 2,500 cases representative of Quebec. *Journal of Occupational Medicine*, **29**, pp. 670-674.

Adams, M.A. and Dolan, P., 1995, Recent advances in lumbar spine mechanics and their clinical significance. *Clinical Biomechanics*, **10**, pp. 3-19.

Battie, M.C. and Bigos, S.J., 1991, Industrial Back Pain Complaints. *The Orthopaedic Clinics of North America*, **22**, pp. 273-282.

Bernard, B.P., 1997, (ed.), Musculoskeletal Disorders and Workplace Factors: A Critical Review of Epidemiologic Evidence for Work-Related Musculoskeletal Disorders of the Neck, Upper Extremity, and Low Back. *National Institute for Occupational Safety and Health* (NIOSH). Publication No. 97-141. Cincinnati.

Bigos, S.J., Battie, M.C., Spengler, D.M., Fisher, L.D., Fordyce, W.E., Hansson, T., *et al.*, 1992, A longitudinal, prospective study of industrial back injury reporting. *Clinical Orthopaedics & Related Research*, **279**, pp. 21-34.

Bongers, P.M., de Winter, C.R., Kompier, M.A.J. and Hildebrandt, V.H., 1993, Psychosocial factors at work and musculoskeletal disease; A review of the literature. *Scandinavian Journal of Work, Environment and Health*, **19**, pp. 297-312.

Bombardier, C., Kerr, M.S., Shannon, H.S. and Frank, J.W., 1994, A guide to interpreting epidemiologic studies on the etiology of back pain. *Spine*, **19**, pp. 2047S-2056S.

Burdorf, A., 1992, Exposure assessment of risk factors for disorders of the back in occupational epidemiology. *Scandinavian Journal of Work, Environment and Health*, **18**, pp. 1-9.

Burdorf, A. and Sorock, G., 1997, Positive and negative evidence on risk factors for back disorders. *Scandinavian Journal of Work, Environment and Health*, **23**, pp. 243-256.

Cohen, J.E., Goel, V., Frank, J.W. and Gibson, E.S., 1994, Predicting risk of back injuries, work absenteeism, and chronic disability. *Journal of Occupational Medicine*, **36**, pp. 1093-1099.

CTD News, 2001 March Issue. *LRP Publications*.

Faucett, J. and Rempel, D., 1994, VDT-related musculoskeletal symptoms: interactions between work posture and psychosocial work factors. *American Journal of Industrial Medicine*, **26**, pp. 597-612.

Frank, J.W., Pulcins, I.R., Kerr, M.S., Shannon, H.S. and Stansfeld, S., 1995, Occupational back pain – an unhelpful polemic. *Scandinavian Journal of Work, Environment and Health*, **21**, pp. 3-14.

Garg, A. and Moore, J., 1992, Epidemiology of low-back pain in industry. *Occupational Medicine: State of the Art Reviews*, **7**, pp. 593-608.

Gibson, E.S., 1988, The value of preplacement screening radiography of the low back. *Occupational Medicine*, **3**, pp. 91-108.

Goldberg, M.S., Scott, S.C. and Mayo, N.E., 2000, A review of the association between cigarette smoking and the development of nonspecific back pain and related outcomes. *Spine*, **25**, pp. 995-1014.

Hadler, N.M., 1986, Regional back pain. *New England Journal of Medicine*, **315**, pp. 1090-1092.

Hoogendoorn, W.E., van Poppel, M.N., Bongers, P.M., Koes, B.W. and Bouter, L.M., 1999, Physical load during work and leisure time as risk factors for back pain. *Scandinavian Journal of Work, Environment and Health*, **25**, pp. 387-403.

Hoogendoorn, W.E., van Poppel, M.N., Bongers, P.M., Koes, B.W. and Bouter, L.M., 2000, Systematic review of psychosocial factors at work and private life as risk factors for back pain. *Spine*, **25**, pp. 2114-2125.

Houtman, I.L.D., Bongers, P.M., Smulders, P. and Kompier, M.A.J., 1994, Psychosocial stressors at work and musculoskeletal problems *Scandinavian Journal of Work, Environment and Health*, **20**, pp. 139-145.

Kerr, M.S., 2000, The importance of psychosocial risk factors in injury. In: Sullivan T.J. *Injury and the New World of Work*. (Vancouver: UBC Press), pp 93-114.

Kerr, M.S., Frank, J.W., Shannon, H.S., Norman, R.W., Wells, R.P., Neumann, W.P. and Bombardier, C., 2001, Ontario Universities Back Pain Study Group. Biomechanical and psychosocial risk factors for LBP at work *American Journal of Public Health*, **91**, pp. 1069-1075.

Kumar S., 2001, Theories of musculoskeletal injury causation. *Ergonomics*, **44**, pp. 17-47.

Liira, J.P., Shannon, H.S., Chambers, L.W. and Haines, T., 1996, Long-term back problems and physical work exposures in the 1990 Ontario Health Survey. *American Journal of Public Health*, **86**, pp. 382-387.

Malchaire, J., Cock, N. and Vergracht, S., 2001, Review of the factors associated with musculoskeletal problems in epidemiological studies. *International Archives of Occupational and Environmental Health*, **74**, pp. 79-90.

Marras, W.S., Davis, K.G., Heaney, C.A., Maronitis, A.B. and Allread, W.G., 2000, The influence of psychosocial stress, gender, and personality on mechanical loading of the lumbar spine. *Spine*, **25**, pp. 3045-3054.

McGill, S.M., 1997, The biomechanics of low back injury: Implications on current practice in industry and the clinic. *Journal of Biomechanics*, **30**, pp. 465-475.

Nachemson, A.L., 1992, Newest knowledge of low back pain – a critical look. *Clinical Orthopaedics and Related Research*, **279**, pp. 8-20.

National Research Council, 2001, Musculoskeletal Disorders and the Workplace: Low Back and Upper Extremities. *Panel on Musculoskeletal Disorders and the Workplace, Commission on Behavioral and Social Sciences and Education, National Research Council.* Washington.

Power, C., Frank, J., Hertzman, C., Schierhout, G. and Li, L., 2001, Predictors of LBP onset in a prospective British study. *American Journal of Public Health*, **91**, pp. 1671-1678.

Rothman, K.J., 1999, *Modern Epidemiology.* (Toronto: Little, Brown), p. 64.

Torp, S., Riise, T. and Moen, B.E., 2001, The impact of psychosocial work factors on musculoskeletal pain: a prospective study. *Journal of Occupational and Environmental Medicine*, **43**, pp. 120-126.

van Tulder, M.W., Assendelft, W.J.J., Koes, B.W. and Bouter, L.M., 1997, Spinal radiographic findings and nonspecific LBP: a systematic review of observational studies. *Spine*, **22**, pp. 427-434.

Volinn, E., Spratt, K.F., Magnusson, M. and Pope, M.H., 2001, The Boeing prospective study and beyond. *Spine*, 26, pp. 1613-1622.

Winkel, J. and Mathiassen, S.E., 1994, Assessment of physical work load in epidemiologic studies: concepts, issues and operational considerations. *Ergonomics,* **37**, pp. 979-988.

Zwerling, C., Ryan, J. and Schootman, M., 1993, A case-control study of risk factors for industrial low back injury. The utility of preplacement screening in defining high-risk groups. *Spine*, **18**, pp. 1242-1247.

CHAPTER 2

Critical Factors in Recovery and Return to Work

Renée-Louise Franche and Niklas Krause

2.1 INTRODUCTION

Return to work (RTW) following an injury or illness is a process which involves the employee, the healthcare providers, the workplace employer/co-workers, and the insurer (Frank *et al.*, 1996). As such, the employee's psychological processes involved in initiating and sustaining RTW cannot be considered in isolation of other elements involved in returning to work. Nevertheless, the employee remains the ultimate agent of change in the RTW process in the sense that only he or she will take the final decision of going in for their day's work.

In this chapter, we set out to capture both the primary agency of the employee as well as the determining impact of interactions with the healthcare system, the workplace, and the insurance system. We accomplish this by combining three models: 1) the Readiness for Change Model (Prochaska *et al.*, 1992) which addresses motivation for behavior change within a social context; 2) the Phase Model of Disability (Krause and Ragland, 1994) which addresses the developmental and temporal aspects of disability; and 3) the model by Frank *et al.* (1996) which outlines the main parties involved in the rehabilitation of employees with occupational health problems.

The Readiness for Change model is of behavioral change which addresses the motivational factors contributing and maintaining behavior change. This model conceptualizes the individual as progressing through stages of change, shifting from the intention to not engage in the given behavior in the foreseeable future, to the intention and ability to perform the given behavior in a sustainable fashion.

The developmental character of disability is increasingly being recognized (Krause and Ragland, 1994; Krause *et al.*, 2001a). Two main models emerge in the literature: an eight phase Occupational Disability Model (Krause and Ragland, 1994) and a three phase model of low back pain (Frank *et al.*, 1998; Spitzer *et al.*, 1987). The eight phase model encompasses two pre-disability phases, i.e. the occurrence of symptoms and the formal report of an injury or illness, and six disability phases. The latter are defined socially by duration of work disability (Krause *et al.*, 2001a). The low back pain phase models delineate three disease phases clinically defined by duration of pain. (Frank *et al.*, 1998; Spitzer *et al.*, 1987). Recent empirical studies building on both models distinguish three main disability phases defined by the number of days off work: acute (up to one month), subacute (two to three months), and chronic (more than three months) (Dasinger *et al.*, 2000; Krause *et al.*, 2001b; McIntosh *et al.*, 2000). Although both models differ in the way they define duration of disability and degree they integrate medical and social aspects of occupational disability, they share important principles: both highlight the phase specificity of risk factors, interactions with the social environment, and interventions. They address the importance of matching occupational and clinical interventions to the appropriate phase of disability.

There is increasing evidence for the phase-specificity of risk factors (Krause *et al.*, 2001). It has been suggested that physical and injury factors are more determining predictors of disability in the acute phase, whereas psychosocial factors have stronger predictive value in the subacute and chronic phases of disability (Shanfield, 1990). The evidence for such a clear-cut distinction of the impact of physical and psychosocial factors is mixed. While the impact of severity of injury has been supported in the acute phase (Krause *et al.*, 2001b; Dasinger *et al.*, 2000; Hogg-Johnson *et al.*, 1994; van der Weide *et al.*, 1999; Dasinger *et al.*, 2001), it continues to be a significant factor for RTW outcomes in the subacute and chronic phases, but to a lesser degree (Krause *et al.*, 2001b; Dasinger *et al.*, 2000). The phase-specificity of healthcare provider recommendation for RTW in the acute phase, and of psychosocial job factors in the subacute/chronic phase (Krause *et al.*, 2001a; Dasinger *et al.*, 2000) have been supported empirically. However, several recent cohort studies (Polatin, 1991) employing phase-specific analyses show that high physical workplace demands are significant predictors of disability throughout all phases (Dasinger *et al.*, 2000; MacKenzie *et al.*, 1998), even after controlling for injury severity and psychosocial job factors (Krause *et al.*, 2001a).

By placing emphasis on the impact of interactions with various parties involved in the recovery/RTW process on the employee's psychological state, we recognize the highly interpersonal nature of the process and give attention to the role of psychological factors within a wider interpersonal context. Getting injured or becoming ill can be traumatic, and the RTW process can also be traumatic if marked by conflict. Consideration of individual psychological factors as determinants of RTW is often met with criticisms invoking such an approach leads to blaming the employee for unsuccessful RTW outcomes. However, an impressive body of evidence exists to support psychosocial and psychological factors as crucial determining elements in the RTW process, in conjunction with physical or pain status (Côté *et al.*, 2001; Hlatky *et al.*, 1986; Linton, 2001; Schade *et al.*, 1999), or when physical/pain status is controlled (Fitzgerald *et al.*, 1989). Also, if psychological and interpersonal factors are not systematically examined, research will be of limited assistance to the psychologically distressed employee.

2.2 THE READINESS FOR CHANGE MODEL

One can conceptualize returning to work as a complex human behavior change, involving physical recovery, motivation, behavior, and interaction with a number of parties. During the last decade, the Readiness for Change Model (Prochaska *et al.*, 1992) has attracted increasing attention as a model addressing both motivational and behavioral factors underlying changes in human behavior, particularly health-risk behaviors. The Readiness for Change Model proposes that relative to a given behavior change, individuals will find themselves in one of five stages. The model suggests that individuals will progress from one stage to the other, however, they can "relapse back" to a previous stage at any point. Five stages are described as they would apply to RTW.

1) *Pre-contemplation*: In this stage, the work-disabled employee is not thinking about initiating behaviors that support adaptive adjustment to his or her RTW. For a severe injury or illness, it may be appropriate for the individual to temporarily put work issues aside in order to focus exclusively on the recovery process.

2) *Contemplation*: The employee is beginning to consider returning to work in the foreseeable future. Although employees are typically engaged in thinking about the pros and cons of returning to work, they are not actively engaged in making concrete plans to do so. The defining characteristic of this stage is ambivalence, where the employee is unable to initiate change because they are stuck in the view that positive benefits of an RTW fail to outweigh negative experiences or outcomes that are also implicated.

3) *Preparation for action*: The employee is making plans to RTW in the near future. Employees are actively seeking information regarding an RTW, testing their abilities to do so, and making a concrete plan to RTW. Employees will be very responsive to help from external sources in order to create an RTW schedule.

4) *Action*: The employee is putting the plan into action and going back to work in some capacity. The employee will continue to be responsive to help from external sources, and motivated to initiate and follow through on targeted behavioral changes. In this stage individuals are at high risk of relapse as they attempt to negotiate their way around potential obstacles. To the extent that they perceive themselves as being successful in returning to work, they will increase their sense of self-efficacy.

5) *Maintenance*: Employees will use specific skills to identify and face high-risk situations that can trigger a relapse back to behaviors that interfere with successful RTW. They will also maintain preventive strategies such as stretching and strengthening exercises for musculoskeletal problems, safety practices, etc.

Three dimensions are involved in mediating progression from Pre-contemplation to Maintenance regarding behavioral change: the decisional balance, self-efficacy, and change processes (see Table 2.1). The decisional balance reflects the cognitive process of weighing the pros and cons of returning to work. As an individual progresses from Pre-contemplation to Maintenance, the decisional balance scale tips as the saliency of the cons of the behavior decreases and the saliency of the pros of the behavior increases. Self-efficacy refers to one's confidence in engaging in RTW and the activities maintaining RTW. It becomes an important factor in Preparation for Action, Action, and Maintenance stages as individuals test their abilities and obtain feedback on their actual ability to RTW. Several studies support the determining role of self-efficacy as a crucial determinant of the likelihood of RTW in individuals (Fitzgerald *et al.*, 1989; Sandstrom and Esbjornsson, 1986). Two categories of processes of change have been described: experiential and behavioral. Experiential processes of change involve change in thoughts, feelings, and attitudes which increase awareness and the perceived need to change, as well as communication with others about the intention or desire to change. These experiential processes are more prominent in the Pre-contemplation and Contemplation stages. Behavioral processes involve actual change in behavior such as increased levels of activity or contacting one's employer, and are more prominent in later stages of the model. Before one engages in behavior changes, one's thoughts, feelings, and attitudes have to be aligned towards RTW.

Table 2.1 Summary dimensions of the stages of change.

Dimension	Pre-contemplation	Contemplation	Preparation For Action	Action	Maintenance
Decisional balance	Cons > Pros	Cons= Pros	Pros > Cons	Pros > Cons	Pros > Cons
Self-efficacy	Low	Low	Moderate	Moderate to high	High
Change processes	Minimal	Experiential	Experiential	Behavioral	Behavioral
General motivational state	Unaware, uninterested	Ambivalent	Committed and motivated to change	Confident Internalizing new behaviors	Internalized behaviors Relapse risk

The model recognizes the contribution of life events as determinants of behavior change, as opposed to relying solely on interventions. Indeed, events such as having children or becoming unemployed can create an emotional arousal conducive to self-reevaluation and behavioral change. The model generalizes to various behaviors (addictive and non-addictive, socially acceptable and not, legal and illegal, frequent and not frequent). It has received impressive empirical support relative to the behaviors of smoking cessation, weight control, delinquency, condom use, sunscreen use, exercise acquisition, mammography screening, and physicians' preventive practices with smokers (Prochaska *et al.*, 1994; Kerns *et al.*, 1997). The model also has excellent predictive validity, particularly with regard to smoking cessation behavior (Velicer *et al.*, 1999). Also, the proposed change processes have been supported by a factor analysis of 770 participants followed-up for six months relative to smoking cessation (Prochaska *et al.*, 1988).

Limitations of the model most pertinent to the area of RTW are the following. First, the model has not been sufficiently validated with non-addictive behaviors (Joseph *et al.*, 1999). Second, study results do not consistently support the presence of five distinct stages. For instance, in a factor analysis relative to pain management behavior, the stage of Preparation for Action was combined in one factor with the Contemplation stage (Kerns *et al.*, 1997). Third, the model poorly addresses the impact of sociodemographic factors, such as age, income, and education, on behavior change, which, in the area of RTW, are of prime importance (Butler *et al.*, 1995).

We now turn to examine how interactions between the injured/ill employee and various parties of the recovery/RTW process can impact on the three dimensions of change – the decisional balance, self-efficacy, and change processes. Although the evidence for the model is strong in relation with health-risk behaviors, no empirical work has been conducted yet examining the Readiness for Change model with regard to the behavior of returning to work following an injury or illness. For that reason, it appears premature to examine in a detailed fashion how the proposed five stages apply to the RTW behavior at this point. However, the three dimensions of change offer a compelling framework to conceptualize the interpersonal impact of the healthcare provider, the workplace, and the insurer on the work-disabled employee.

2.3 INTERACTING WITH THE MAIN PLAYERS IN THE RECOVERY AND RTW PROCESS

2.3.1 The Healthcare Provider

Healthcare providers have the potential to become trusted sources of information by the injured employee, depending on the degree of independence from the employee's insurer or workplace. They can foster realistic or unrealistic expectations in an employee regarding the course, nature, and speed of recovery and their ability to RTW. The impact of the healthcare provider on the employee's readiness for change needs to be considered within the framework of the developmental models of disability.

2.3.2 Decisional Balance

Interactions with healthcare providers may be particularly potent as determinants of pros and cons of RTW. Messages of healthcare providers concerning factors influencing health, including RTW, work environment and type of work offered, will bear considerable weight in the injured employee's decisional balance. Direct physician advice to RTW has an impact on RTW rates. A retrospective study of 325 claimants in California with a 3.7 years follow-up period supports the important role of the doctor's recommendation to RTW in a phase-specific way: the positive recommendation for readiness for RTW was associated with a 39% higher RTW rate during the acute disability phase and a 67% higher RTW rate during the subacute and chronic phases (Dasinger *et al.*, 2001). After adjustment for possible confounders, including patient demographics, injury severity, previous injuries, physical and psychosocial job factors, length of time at the pre-injury job, and employer size, the effect of the doctor's advice to RTW was attenuated to a statistically non-significant 24% increase in RTW rates during the acute phase and an only marginally statistically significant but still substantial increase of 42% during the subacute/chronic disability phase. The smaller effect in the acute phase of the recovery suggests that severity of injury and physical and psychosocial job characteristics may dominate one's ability to RTW during the acute phase and limit the doctor's direct influence on duration of work disability.

The same study examined employees' recollection of their interaction with their doctor in terms of the doctor's proactive efforts to provide information about work restrictions and modifications, and to recommend employee behavior facilitating recovery and prevention of future injury (Dasinger *et al.*, 2001). While doctor-proactive communication regarding RTW was associated with a higher likelihood to RTW in the acute phase of recovery, this effect was reduced by half and was no longer statistically significant when physical and psychosocial workload factors were added in the statistical model. There was no effect of doctor's proactive communication in the subacute and chronic phases. Taken together, these results suggest that the phase-specific impact of the healthcare provider as a facilitator of RTW remains limited if it does not translate into actual changes in the physical and psychosocial workload of the employee. It points to the importance of the healthcare provider not solely conveying work-related information to the employee but also liaising with the workplace to ensure that appropriate work accommodations are in fact available.

In patients having suffered an uncomplicated myocardial infarction, a specific recommendation by a physician to RTW early on resulted in earlier RTW, lower perception of disability, and increased productivity at six month post-cardiac event, as opposed to a group who did not receive this intervention. At the one year follow-up, RTW rates and hours worked per week were however similar (Dennis *et al.*, 1988). In another intervention study, the impact of insurance medical advisors' rehabilitation-oriented program was compared with "regular care" (Donceel and Du Bois, 1999). The program included, among other components, a clear statement to the patient about the expected duration of work incapacity. The program as a whole resulted in a significantly higher RTW rate one year post-surgery, however the independent contribution of the physician's communication of expected duration of work incapacity was not assessed.

Overall, studies suggest direct advice from a healthcare provider can have an impact on RTW rates, but it needs to be examined within the larger context of severity of injury, sociodemographic factors, and workplace factors.

2.3.3 SELF-EFFICACY

Expectations regarding recovery will have an impact on one's sense of self-efficacy to engage in many activities, including RTW. A recent review of 16 high-quality studies examining the impact of expectations about recovery from a wide variety of physical and psychiatric conditions (Mondloch *et al.*, 1999) supports the important role of patient expectations on outcomes such as subsequent subjective well-being after minor surgery (Flood *et al.*, 1993), physical functional ability after a cardiac event (Allen *et al.*, 1990; Diederiks *et al.*, 1983; Petrie *et al.*, 1996; Ruiz *et al.*, 1992), and psychological adjustment following surgery (Jamison *et al.*, 1987). Regarding the specific outcome of RTW, the review found that in patients having had a myocardial infarction, recovery expectations were predictive of their RTW rate at six weeks (Petrie *et al.*, 1996), six months (Maeland and Havik, 1987), and one year post-cardiac event (Diederiks *et al.*, 1983). In employees with occupational musculoskeletal disorders, while one study did not support the predictive value of initial RTW expectations (Pransky *et al.*, 1999), in a small sample of patients with low back pain, the expectation of not being able to "manage" returning to work was associated with a higher likelihood of not being back at work or in retraining one year post-injury, and a higher likelihood of being "sick-listed" for more than six months, four years post-injury (Sandstrom and Esbjornsson, 1986). As well, in the Early Claimant Cohort study (Cole and Mondloch, 1999), a prospective study involving 1,332 Canadian employees with lost-time claims following an injury, expectations about recovery measured within three weeks of their reported injury were statistically significantly associated with longer periods receiving benefits within the first year following injury, even when controlling for pain level, quality of life, functional status, co-morbidities, sex, income, marital status, age, job demands, and workplace accommodations.

Self-efficacy is decreased by depressive symptomatology (Norman *et al.*, 1997). The impact of depression, mediated by self-efficacy, may decrease the likelihood of engaging in successful change processes regarding RTW. Indeed, the evidence supporting the impact of depressive symptomatology on RTW outcomes is substantial. In a prospective study of 46 patients with lumbar discectomy, RTW two years post-surgery was predicted by depression and occupational mental stress, and not by clinical findings or morphological alterations identified with magnetic resonance (Schade *et al.*, 1999). Similarly, injured employees on compensation benefits with moderate to severe depressive symptomatology were significantly less likely to RTW following vocational rehabilitation than individuals with lower depressive score (Ash and Goldstein, 1995). In a sample of 7,462 individuals with whiplash followed prospectively over one year, the presence of depressive symptomatology was associated with a 37% reduction of claim-closure rate under a tort system and a 36% reduction under a no-fault system (Côté *et al.*, 2001).

The role of the healthcare provider in monitoring, diagnosing, and treating depression is crucial. Unfortunately, depression is underdiagnosed in primary care settings: 35% to 70% of primary care patients with depression do not receive a diagnosis or receive inadequate treatment (Depression Guideline Panel, 1993; Coyne *et al.*, 1994; Hirschfeld *et al.*, 1997). The extent to which the same underdiagnosis and undertreatment is occurring in work-disabled individuals remains unknown. It is noteworthy that anecdotally, depressive symptomatology is often ignored by healthcare providers and other parties involved in the RTW process in an effort to avoid "blaming the worker". It is clearly advantageous to attend to the symptoms of depression in the work-disabled employee.

Even when RTW self-efficacy is measured early in the recovery process, it has a significant impact on RTW rates. Indeed, when measured prior to hospital discharge in patients having suffered a cardiac event, RTW self-efficacy was the strongest predictor of six month self-reported full-time or part-time RTW, independently of having a previous myocardial infarction, disease severity, age, job classification, and gender (Fitzgerald *et al.*, 1989). This underscores the importance of increasing an employee's self-efficacy early in the recovery process.

2.3.4 Change Processes

When a person enters the long-term disability phase as defined by the Phase Model of Disability (Krause and Ragland, 1994), around eight weeks post-injury, the healthcare provider will again play a determining role in coordinating clinical interventions (Loisel *et al.*, 1994). More intense rehabilitation efforts may be beneficial at that point and iatrogenic effects are less likely to occur, as in the acute phase of disability (Sinclair *et al.*, 1997). Healthcare providers also play a key role in the decision about the employee's RTW readiness; without a physician's clear signal that an employee can safely RTW, no work accommodation can be initiated.

One recent study highlights the importance of integrating both healthcare provider care and workplace accommodation process (Loisel *et al.*, 1997): in the city of Sherbrooke, Canada, 130 employees from 31 workplaces who had been absent from work for more than four weeks for back pain, were randomized, based on their workplace, in one of four treatment groups: usual care, clinical intervention, workplace ergonomic intervention, and full intervention (a combination of the last two). The full intervention group returned to regular work 2.41 times faster than the usual care intervention group, supporting the benefits of integrating clinical and ergonomic interventions. Clinical intervention alone had no beneficial effects over usual care. Workplace ergonomic intervention appeared to be responsible for most of the beneficial effects observed in a combined approach.

2.3.5 How Can the Healthcare Provider Facilitate the Employee's Progress in Readiness to RTW?

The healthcare provider may help resolve an employee's ambivalence about returning to work by raising the salience of the pros of a safe and early RTW. A respectful discussion of the beneficial role of a gradual and paced return to daily and occupational activities, in preventing prolonged disability, may begin to shift the decisional balance. Reassurance regarding the normalcy of certain symptoms may decrease employees' anxiety about their health status and about the safety of an RTW. Accurate information regarding the appropriate use of medication can also decrease one's fear of pain and fear that it signals further physical damage.

The evidence supporting the impact of a doctor's recommendation to RTW and of patient self-efficacy and expectations on the RTW process is compelling. It is likely that the healthcare provider can significantly increase an employee's self-efficacy regarding RTW by providing factual information about the probability of recovery and by encouraging self-management of symptoms when appropriate. Psychological distress requires careful monitoring and clinical attention in order to optimize a fair and effective RTW process. Also, coordination of healthcare and ergonomic interventions appear to be a key element of increasing readiness to RTW for individuals. It is also clear that healthcare provider impact needs to be considered within the context of other potent factors, such as past experiences with injuries or illnesses, workplace and social factors.

2.3.6 The Worksite

Workplace disability management strategies are repeatedly found to be critical and determining factors of RTW outcomes (Hunt *et al.*, 1993; Hogg-Johnson and Cole, 1998; Krause *et al.*, 1998a). Disability management can be described as a "proactive, employer-based approach developed to a) prevent the occurrence of accidents and disability b) provide early intervention services for health and disability risk factors, and c) foster coordinated administrative and rehabilitative strategies to promote cost-effective restoration and RTW" (Habeck *et al.*, 1991). We will focus primarily on RTW strategies and will differentiate structure from process in disability management.

We define structure as referring to the specific programmatic components of disability management, as opposed to process, which refers to the manner in which these components are offered. Initial studies in the area of RTW tended to focus on the presence or absence of structure of disability management (Hunt *et al.*, 1993). These studies examined how strategies such as having an RTW coordinator, provision of work accommodations, and monitoring of outcomes, impacted on RTW outcomes such as claim rates. During the last few years, more attention has been devoted to the process of RTW strategies. Qualitative (Clarke *et al.*, 2000; Côté *et al.*, 2000) and empirical studies (van der Weide, 1999) have repeatedly underscored the importance of interactional factors in the process of RTW programs and these will be highlighted in the following section as prognostic factors of RTW after the occurrence of an injury.

2.3.7 Decisional Balance

Interactions with the worksite will greatly affect one's decisional balance of the pros and cons of returning to work. Of primary concern are the employer's and co-workers' responses, the safety of the job returned to, and factors precipitating a premature RTW. Qualitative studies of injured employees suggest that a non-confrontational and non-judgmental approach from the worksite after injury is essential to successful RTW (Clarke *et al.*, 2000; Côté *et al.*, 2000). Empirical studies generally provide evidence supporting the importance of the employer's response. In a study of employees with work-related upper extremity disorders, employees who were work-disabled indicated being significantly more angry toward their employer as compared to employees who returned to work or never left work following the injury (Himmelstein *et al.*, 1995). In a study of 120 Dutch workers with work absences of 10 days or more due to low back pain, problematic relations with colleagues was one of four predictive factors of time off work (van der Weide *et al.*, 1999).

One concept which has received increased attention is the one of legitimacy (Tarasuk and Eakin, 1995; Trasuk and Eakin, 1994; Smith *et al.*, 1998). Legitimacy refers to the degree to which an injured employee feels believed by others regarding the authenticity of their injury and of their symptoms. It is of particular relevance to injuries and illnesses which involve work absences and to injuries/illnesses which are "invisible", such as soft tissue sprains and strains. In the Early Claimant Cohort study, legitimacy was a significant predictor of duration on benefits (Smith *et al.*, 1996). The mechanisms underlying the association between legitimacy and RTW outcomes remain speculative. If an employee feels workplace staff question the legitimacy of symptoms, the worker may develop negative feelings towards the workplace which will weigh against returning to work in their decisional balance. Alternatively, the perceived expression of disbelief of one's symptoms and complaints may bring an employee to invest energy in "proving" their injury and pain are real by not returning to work.

Workplace culture, which refers to a general interpersonal and value-focused atmosphere, is also associated with RTW outcomes (Hunt *et al.*, 1993). In one prospective study of 198 employees with carpal tunnel syndrome, an increased level of people-oriented culture in the workplace was associated with higher RTW rates six months after cases had been identified in community medical practices, when age, gender and baseline carpal tunnel syndrome symptom severity were controlled for (Amick *et al.*, 2000). People-oriented culture was a factorially derived dimension defined as "the extent to which the company involves employees in meaningful decision making, where there is trust between management and employees, and openness to share information in a cooperative work environment." (p. 30).

If work is regarded as a threat to one's health, clearly this will weigh heavily in weighing the pros and cons of returning to work. Fear/avoidance constructs have been examined as they relate to physical activity and work – that is the degree to which an individual fears and/or avoids physical activity and/or work as a result of being concerned about the impact of physical activity/work on symptoms. In one study of 63 individuals with low back pain, anticipation of pain and fear-avoidance beliefs about physical activities were the strongest predictors of variations in physical performance (Al-Obaidi *et al.*, 2001). In a cross-sectional study of 184 employees with back pain, the best predictor of RTW was the employee's fear-avoidance beliefs concerning the impact of RTW on their symptoms and their physical state: fear-avoidance belief about work accounted for 26% of the variance for prior work loss, and 23% of the variance of disability, even after controlling for severity of pain (Waddell *et al.*, 1993).

The goal of RTW programs is generally to achieve an early and safe RTW. Risk factors contributing to a premature and unsafe RTW are seldom considered. It is important to consider that fear of losing one's job and financial strain will weigh in one's decision balance and can contribute to the decision to go back to work too soon, increasing risk of re-injury and long-term ill health (Pransky *et al.*, 2000).

2.3.8 Self-efficacy and Change Processes

Returning to work gradually in terms of hours worked or with modified work can have a significant impact on one's self-efficacy regarding one's ability to return to original work. Through the mediation of increased self-efficacy, the offer and acceptance of a modified work accommodation may represent the decisive change process for successful RTW (Hunt *et al.*, 1993; Krause *et al.*, 1998b).

In a systematic review of the scientific literature on modified work from 1975 to 1997, Krause and colleagues identified 13 high-quality studies out of 29 empirical studies (Krause *et al.*, 1998a). Based on the high-quality studies, they concluded that injured employees who are offered modified work are twice as likely to RTW than those who are not offered such an arrangement. Also, modified work reduces by half the number of lost days. Satisfactory work accommodations have been shown to significantly decrease injured employees' anxiety about returning to work (Pransky *et al.*, 2000). Offers of modified work also reduce the duration of receiving compensation benefits (Bernacki *et al.*, 2000; Hogg-Johnson and Cole, 1998), and reduce the incidence of injury in general (Amick *et al.*, 2000; Gice and Tompkins, 1989; Yassi *et al.*, 1995).

Although modified work can speed up RTW, insufficient attention to risk factors can lead to re-injury. Indeed, attention should be given to the danger of returning the employee to work too early. The physician who wishes to avoid the vicious cycle of inactivity and deconditioning may in fact return an employee prematurely. The employee may still not have the physical strength to sustain the proposed work program. It is generally accepted that an employee does not need to wait until he or she is "100%" to initiate an RTW. But what if it is too soon? How can it be accurately determined? These questions remain some of the most interesting and relevant ones in the field of rehabilitation.

The interpersonal aspects of modified work are numerous. The importance of a supportive supervisor response has been raised in qualitative studies (Frank and Guzman, 1999) and in empirical studies (Krause *et al.*, 2001a; Bergquist-Ullman and Larsson, 1977; Krause *et al.*, 1997; Krause *et al.*, 2001b). In a study of 434 employees with low back pain, low supervisor support reduced RTW rates by 21% (Krause *et al.*, 2001a). The role of co-worker support has received mixed results. In one study, it was found to be associated with longer duration of work disability (Bergquist-Ullman and Larsson, 1977), while in other studies it had no effect (Krause *et al.*, 1998b; Krause *et al.*, 2001a). The inconsistency of results may be related to the various forms of "co-worker support": co-workers may provide support by cooperating with the injured employee in their modified work programs, or they may also support the injured employee in their pain/limited behavior by suggesting that the injured employee take more time off out of concern for the injured employee. Gender may also affect the impact of relationships with co-workers: in one British study, a trend towards greater effects of co-worker support in women than in men was found (Papageorgiou *et al.*, 1997). Nevertheless, co-workers' concerns need to be considered when planning a modified work arrangement: co-workers whose safety is put at risk, or who experience an increase in workload as a result of a modified work accommodation may not welcome or facilitate the arrangement. Self-efficacy is not only determined by one's ability to be successful at the given modified work but also by one's sense of value derived from the modified work. Clinicians and researchers are becoming acutely aware that meaningless and demeaning modified work, which does not contribute to the overall functioning of one's workplace, can actually do more harm than good in the RTW process (Clarke *et al.*, 2000).

Other factors impacting on the probability of success of a modified work accommodation and on self-efficacy deserve mention although they are less interpersonal in nature. The pace and nature of modified work and the ergonomic aspects of work (Amick *et al.*, 2000; Adams *et al.*, 1994; National Institute for Occupational Safety and Health, 1997; Nachemson, 1983; Ranney, 1997) have a significant impact on RTW outcomes. Size (Eakin *et al.*, 2000; Habeck *et al.*, 1991; Cheadle *et al.*, 1994; Hunt *et al.*, 1993; Dasinger *et al.*, 2000) and sector of workplaces (McIntosh *et al.*, 2000; Habeck *et al.*, 1991) will also impact on the range of modified work accommodations available, as well as on the general culture of workplaces (Habeck *et al.*, 1991).

2.3.9 Future Directions for Workplace Interventions

Much research on the impact of disability management strategies has thus far focused on determining the structural elements critical to effective RTW programs. Some studies did not evaluate the specific contribution of individual elements of their interventions, in which a number of strategies were simultaneously evaluated (Donceel and Du Bois, 1999). Other studies attempted to separate the contribution of various rehabilitation components and pointed to the greater importance of ergonomic factors and modified work accommodations, when compared to clinical (physical and psychological) interventions (Loisel *et al.*, 1997). More recent literature is now pointing to the importance of considering process or interpersonal aspects of RTW strategies. Workplace culture, as well as supervisors' interactions with injured and ill workers are now recognized as influential factors of RTW strategies. The degree to which interactions with workplace staff impact on the pros and cons of returning to work remains to be determined. But a hostile environment will decrease the perceived advantages of returning to work.

Of concern is the realization that workplace offers of modified work, and interpersonal processes surrounding the presence or absence of a modified work offer, are often less than optimal. Indeed, in the Ontario Early Claimant Cohort study (Hogg-Johnson and Cole, 1998), 74% of the sample interviewed who were interviewed within three weeks after an accepted claim reported negative supervisor response to their injury. In that same cohort, 73% reported that there had not been any offers of modified work.

A collaborative and respectful approach from workplace parties will clearly lead to a climate of trust much more conducive to reducing one's anxieties about returning to work and to shifting the decisional balance towards being ready for an RTW attempt. Attention to co-workers' safety, workload and understanding of modified work will also set the stage for a modified work accommodation conducive to increased self-efficacy and long-term success in RTW.

2.3.10 The Insurer

Available evidence suggests that most appeals of disability claims are brought on by the consequences of the disability, rather than the appeal process contributing to the development of disability. Indeed, an American study (Volinn *et al.*, 1991) found that appeals of disability claims were consistently made long after the date of injury – none were made within the first 90 days, and only one-fourth were made within the first year post-injury. With regards to establishing the direction of causality, this study strongly suggests that it is the disability and its consequences which brings individuals to appeal their disability claims. Moreover, anecdotal information suggests that litigation in workers' compensation in the United States is most often initiated by the employer and not by the employee, possibly as a result of a routine internal policy to discourage claimants from pursuing benefits, and to decrease insurer cost.

The impact of the insurance provider and of types of benefits available on RTW has been the focus of ample discussion and controversy. The literature has focused on the financial aspects of available compensation and on the impact of litigation, which pertains mostly to the decisional balance aspect of readiness for RTW. For that reason, only the decisional balance will be discussed within the context of the role of the insurer, and not the self-efficacy and change processes.

2.3.11 The Employee's Decisional Balance

The impact of compensation

Before one considers the impact of benefits on RTW, it is important to step back and appreciate the fundamental purpose of disability compensation systems. "The primary purpose of workers' compensation is to help employees who sustain on-the-job injuries recover and RTW and/or be compensated by any resulting permanent disability" (Wilkinson, 1994, p. 28) without recourse to potentially costly legal action. The refusal of such needed financial assistance can result in the denial of social legitimacy (Wilkinson, 1994), which in turn can lead to anxiety and depressive affect (Ison, 1986). The profound negative consequences for the employee of a wrongfully denied claim have been recognized in cases where compensation was granted not just to the initial work injury but for the disability (most often depression) related to the compensation process itself in Canada (Lippel, 1999) and the United States (Kornblum, 1979).

Financial gains associated with disability benefits for lost time can factor in the decisional balance. There is substantial controversy over the magnitude and even the existence of the impact of such financial gains. While some studies found no effect of compensation on RTW rates (Hogelund, 2000; Waehrer *et al.*, 1998; Yelin, 1986), other studies found that increases in benefits are associated with increases in frequency and duration of claims (Fortin and Lanoie, 1992; Lanoie, 1992; Dionne *et al.*, 1995; Kralj, 1995; Johnson *et al.*, 1995; Thomason and Pozzebon, 1995). These studies are limited in that they do not focus on actual RTW, but on duration of claims only. Two studies of Canadian workers from the province of Ontario did focus on actual RTW and found that higher benefit rates were associated with lower rates of RTW (Butler *et al.*, 1995; Hyatt, 1996).

Other studies have considered the impact of compensation on employees' RTW rates as it interacts with other factors. These studies generally point to a more complex situation, in which compensation does not unequivocally translate to lower RTW rates. One multinational prospective study, the Work Incapacity and Reintegration Project (Bloch and Prins, 2001) compared the effectiveness of different RTW interventions used by social security systems and healthcare providers in six countries (Denmark, Germany, Israel, the Netherlands, Sweden, and the USA). Homogeneous cohorts of employees who were work-disabled for at least three months because of low back disorders were followed over a two year period, with data collection at baseline and one and two years follow-up. The association between wage replacement benefits and duration of disability was dependent on the degree of job security. The combination of extensive benefits with strong job protection predicted early RTW, while weak job protection combined with extensive benefits did not improve RTW. Shorter disability periods were seen for low levels of job protection combined with low levels of benefits, but mainly for new employees. Clearly, individuals weigh the pros and cons associated with their given levels of wage replacement and job security.

A critical American study of 312 individuals with severe lower extremity fractures, found that receiving disability benefits had a strong negative effect on RTW during the initial six months post-injury only (MacKenzie *et al.*, 1998). This effect was particularly apparent during the first three months of recovery, which corresponds to the typical necessary time to heal for a fracture. At six to twelve months post-injury, there were no statistical differences between those receiving compensation and those who did not. While this study points to the impact of compensation, it also points to its complexities. It brings to the forefront the fact that individuals receiving no compensation may RTW too soon due to financial pressures and jeopardize their health. As the authors of that study state: "It is.... important to acknowledge, however, that compensation often affords the opportunity for more complete recovery and minimizes the potential for re-injury, subsequent work-loss days, or failure at the job owing to residual limitations not yet resolved at the time of reentry into the workforce" (p. 1635). The study also points again to the importance of examining the role of determinants of RTW through the lens of the phase-specific approach.

The role of the family in the decisional balance needs to be discussed in terms of its impact on financial needs. Each family member is either a dependent, an income provider, or both. As such, they represent a stressor or buffer on financial strain. One American study found an interaction between wage replacement and the number of children combined with marital status, with length of work disability as the outcome (Volinn *et al.*, 1991): given equal wage compensation ratios, widowed and divorced individuals were twice as likely as single individuals to be work-disabled for 90 days or more. This effect was lessened by the presence of children for widowed and divorced individuals. Also, when one considers significant predictors of disability retirement in a Finnish population, increased number of family members working and fewer unemployed family members are associated with increased likelihood of a disability retirement (Krause *et al.*, 1997). Overall, when assessing the influence of compensation on RTW, the above studies highlight the importance of analyzing it in conjunction with other factors, such as job security, ratio of income replacement to previous earnings, amount of regular income, number of dependents, and financial pressure to RTW. When these factors are taken into account, the impact of compensation in the decisional balance is less absolute and invariant than was once believed.

Administration of benefits

The manner in which compensation is administered will also factor in the decisional balance. If workers are concerned about their ability to perform full-time work and face the prospect of losing all compensation if they return only part-time, despite their desire to RTW they may opt to delay the RTW (Ferrier and Lavis, 2001). This points to a major weakness in most disability income support programs, that is, the dichotomization of employability. Individuals are categorized as employable or not employable, which has a direct impact on their eligibility for benefits. This polarized view of employability does not acknowledge that some individuals may be able to work only with accommodations, removal of barriers, adapted transportation, or on a part-time basis only.

The lack of coordination between different compensation agencies can have quite different effects on the decisional balance of the worker. On the one hand, it can lead to cost saving efforts by shifting claimants to other systems, minimizing their chances of receiving adequate compensation and increasing their level of frustration. On the other hand, lack of coordination can lead to a situation where claimants receive more than they were earning before the injury due to multiple insurance sources (disability compensation, mortgage insurance, etc.).

2.4 MODEL APPLICATIONS

The relevance and potential applications of the Readiness for Change model for the RTW process have been reviewed in this chapter, with a particular focus on the three dimensions of change – the decisional balance, self-efficacy, and change processes. The limitations, strengths and associated future directions for research relative to the Readiness for Change model will now be discussed.

The Readiness for Change model places the injured/ill employee as the primary agent of change, as he or she interacts with various parties in the RTW process. As such it does not comment on the interpersonal impact of the employee on the employer, healthcare provider, and insurer, but focuses solely on the unidirectional impact of the latter parties on the employee. This model provides a solid framework for the first step in the study of the interpersonal aspects of the RTW process, but to fully capture its dynamic process, one would need to go one step further and consider the impact of the employee on other parties. The role of co-workers and unions need to be considered as they are important forces facilitating an early and safe RTW.

The Readiness for Change model posits specific timeframes to consider when attempting to "stage" an injury claimant (determining in which stage a person is in). These timeframes may not apply to the behavior of returning to work following an injury, and may vary depending on the degree of severity and the type of the injury or illness. The more severe the impairment, the longer one would be expected to remain in the Pre-contemplation and Contemplation stages regarding RTW, as one's energies are focused on physical recovery. Future research will clarify appropriate timeframes for each stage. The Readiness for Change model posits that, depending on the stage of change of an individual, the effectiveness of offered interventions will vary immensely. In that sense, prescribed interventions are stage-based, targeting the dimensions of readiness most likely to be modified. This approach offers a good fit with recent modeling of optimal RTW interventions which emphasize stage-specificity of risk factors and interventions (Krause and Ragland, 1994; Loisel *et al.*, 1994; Loisel *et al.*, 1997).

The Phase Model of Occupational Disability provides the conceptual and analytical framework for epidemiological research that takes into account the developmental nature of the disability process and its reversal during recovery and RTW. Recent epidemiological studies supported the presence of specific predictors of disability during the acute, subacute, and – to a lesser extent – chronic phases of disability. The Phase model is instrumental in identifying and discriminating predictors of disability that are influential only in certain phases of disability, throughout all phases of disability, or change their impact across disability phases. It is important to note that effect estimates of these phase-specific predictors reflect the average experience of all individuals in the underlying population and do not clarify at which exact time an intervention should be offered to any specific individual within each disability phase. The Readiness for Change model has the ability to account for individual variation in optimal timing of interventions, based on an individual's perceived readiness for RTW. The Readiness for Change model therefore potentially complements the Phase Model of Occupational Disability by allowing for an individual-level staging of the disability and recovery process within the broader (group-level derived) occupational disability phases framework.

More research is required to achieve a better understanding of how the Phase Model of Disability and the Readiness for Change model interface. One step to explore the match between the two models would be to assess the distribution of stages of individuals within one phase of disability. When individuals are in the subacute phase of disability, how many are still in the Pre-contemplation stage, the Contemplation phase, or have moved to the Preparation phase or Action phase? Another important application of the Readiness for Change model will be to use the staging of individuals as predictors of their progression, or absence of progression, to the subacute and chronic phases of disability. An example would be to examine whether staging of individuals at the beginning of a work accommodation program could be predictive of the sustainability of their RTW. Although the Readiness for Change model is solidly evidenced-based regarding health-risk behaviors, it remains a heuristic theoretical model regarding its application for RTW behavior. Future efforts should focus on empirically testing its applicability and predictive power in injured and ill work-disabled employees.

2.5 KEY MESSAGES

- The Readiness for RTW model and the Phase Model of Disability both highlight the dynamic and evolving nature of the disability process, and conversely of recovery. They acknowledge the interactive process taking place between employee, employer, healthcare provider, and insurer, and its interaction with time since the injury of illness.

- Healthcare providers' communication with employees is an important influence on employees' expectations and decision making process related to returning to work.

- To optimize RTW outcomes, clinical and workplace interventions require coordination and integration.

- The critical element of successful RTW programs is a work accommodation offered in an interpersonally positive climate.

- It is essential to begin to consider compensation as a critical and necessary element for safe RTW. Adequate compensation, as well as job security, decrease undue financial pressure to RTW and allow appropriate time for the healing and recovery processes to take place.

2.6 ACKNOWLEDGEMENTS

We wish to thank Dr Curtis Breslin from the Institute for Work & Health for his insightful comments on a previous draft of this chapter.

2.7 REFERENCES

Adams, M.L., Franklin, G.M. and Barnhart, S., 1994, Outcome of carpal tunnel surgery in Washington State Workers' Compensation. *American Journal of Industrial Medicine*, **25**, pp. 527-536.

Allen, J.K., Becker, D.M. and Swank, R.T.,1990, Factors related to functional status after coronary artery bypass surgery. *Heart & Lung*, **19**, pp. 337-343.

Al-Obaidi, S.M., Nelson, R.M., Al-Awadhi, S. and Al-Shuwaie, N., 2001, The role of anticipation and fear of pain in the persistence of avoidance behavior in patients with chronic low back pain. *Spine*, **25**, pp. 1126-1131.

Amick, B.C., Habeck, R.V., Hunt, A., Fossel, A.H., Chapin, A., Keller, R.B. and

Katz, J.N., 2000, Measuring the impact of organizational behaviors on work disability prevention and management. *Journal of Occupational Rehabilitation*, **10**, pp. 21-38.

Ash, P. and Goldstein, S.I., 1995, Predictor of returning to work. *Bulletin of American Academy of Psychiatry Law*, **23**, pp. 205-210.

Bergquist-Ullman, M. and Larsson, U., 1977, Acute low back pain in industry. A controlled prospective study with special reference to therapy and confounding factors. *Acta Orthopaedica Scandinavia* (Suppl), **170**, pp. 1-117.

Bernacki, E.J., Guidera, J.A., Schaefer, J.A. and Tsai, S., 2000, A facilitated early RTW program at a large urban medical center. *Journal of Occupational and Environmental Health*, **42**, pp. 1172-1177.

Bloch, F. S. and Prins, R. (eds), 2001, *Who returns to work and why? A six-country on work incapacity and reintegration.* (New Brunswick, New Jersey: Transaction Publishers).

Butler, R.J., Johnson, W.G. and Baldwin, M., 1995, Managing work disability: Why first RTW is not a measure of success. *Industrial Labor Relations Review*, **48**, pp. 452-469.

Cheadle, A., Franklin, G., Wolfhagen, C., Savarino, J., Liu, P.Y., Salley, C., Py, L. and Weaver, M., 1994, Factors influencing the duration of work-related disability: a population-based study of Washington state workers' compensation. *American Journal of Public Health*, **84**, pp. 190-196.

Clarke, J., Cole, D. and Ferrier, S., 2000, *RTW after a soft tissue injury: A qualitative exploration.* Working Paper #127. Institute for Work & Health. Ref Type: Unpublished Work.

Cole, D.C. and Mondloch, M.V., 1999, ECC Prognostic Modelling Group. *Listening to workers: how recovery expectations predict outcomes.* Working Paper #73. (Toronto: Institute for Work & Health), pp. 1-29.

Côté, P., Clarke, J., Deguire, S., Frank, J.W. and Yassi, A., 2000, *A report on chiropractors and RTW: the experiences of three Canadian focus groups.* (Toronto: Institute for Work & Health), pp. 1-23.

Côté, P., Hogg-Johnson, S., Cassidy, J.D., Carroll L. and Frank, J.W., 2001, The association between neck pain intensity, physical functioning, depressive symptomatology and time-to-claim-closure after whiplash. *Journal of Clinical Epidemiology*, **54**, pp. 275-286.

Coyne, J.C., Fechner-Bates, S. and Schwenk, T.L., 1994, Prevalence, nature, and comorbidity of depressive disorders in primary care. *General Hospital Psychiatry*, **16**, pp. 267-276.

Dasinger, L.K., Krause, N., Deegan, L.J., Brand, R.J. and Rudolph, L., 2000, Physical workplace factors and RTW after compensated low back injury: a disability phase-specific analysis. *Journal of Occupational and Environmental Medicine*, **42**, pp. 323-333.

Dasinger, L.K., Krause, N., Thompson, P.J., Brand, R.J. and Rudolph, L., 2001, Doctor proactive communication, RTW recommendation, and duration of disability after a workers' compensation low back injury. *Journal of Occupational and Environment Medicine*, **43**, pp. 515-525.

Dennis, C., Houston-Miller, N., Schwartz, R.G., Ahn, D.K., Kraemer, H.C., Gossard, D., Juneau, M., Taylor, C.B. and DeBusk, R.F., 1988, Early RTW after uncomplicated myocardial infarction. *Journal of the American Medical Association*, **260**, pp. 214-220.

Depression Guideline Panel – U.S. Department of Health and Human Services, Public Health Service Agency for Health Care Policy and Research, 1993. *Depression in Primary Care.* Volume 1. *Detection and Diagnosis. Clinical Practice Guideline,* no. 5. AHCPR publication no. 93-0550. 1993. Rockville, MD. Ref Type: Report.

Diederiks, J.P.M., van der Sluijs H., Weeda, H.W.H. and Schobre, M.G., 1983, Predictors of physical activity one year after myocardial infarction. *Scandinavian Journal of Rehabilitation Medicine*, **15**, pp. 103-107.

Dionne, C., Koepsell, T.D., Von Korff, M., Deyo, R.A., Barlow, W.E. and Checkoway, H., 1995, Formal education and back-related disability: in search of an explanation. *Spine*, **20**, pp. 2721-2730.

Donceel, P. and Du Bois, M., 1999, Predictors of work incapacity continuing after disc surgery. *Scandinavian Journal of Work, Environment and Health*, **25**, 264-271.

Eakin, J.M., Lamm, F. and Limborg, H.J., 2000, International perspective on the promotion of health and safety in small workplaces, In: Frick, K., Jensen, P.L., Quinlan, M. and Wilthagen. T., editors. *Systematic Occupational Health and Safety Management*. (New York: Pergamon) pp. 227-247.

Ferrier, S. and Lavis, J., 2001, *With Health Comes Work?* (Toronto: Institute for Work & Health), Ref Type: Report.

Fitzgerald, S., Becker, D., Celentano, D., Swank, R. and Brinker, J., 1989, RTW after percutaneous transluminal coronary angioplasty. *American Journal of Cardiology*, **68**, pp. 1108-1112.

Flood, A.B., Lorence, D.P., Ding, J., McPherson, K. and Black, N.A., 1993, The role of expectations in patients' reports of post-operative outcomes and improvement following therapy. *Medical Care*, **31**, pp. 1043-1056.

Fortin, B. and Lanoie, P., 1992, Substitution between unemployment insurance and workers' compensation: an analysis applied to the risk of workplace accidents *Journal of Public Economics*, **49**, pp. 287-312.

Frank, J., Sinclair, S., Hogg-Johnson, S., Shannon, H., Bombardier, C., Beaton, D. and Cole, D., 1998, Preventing disability from work-related low-back pain: new evidence gives new hope – if we can just get all the players on onside. *Canadian Medical Association Journal,* **158**, pp. 1625-1631.

Frank, J.W., Brooker, A.S., DeMaio, S., Kerr, M.S., Maetzel, A., Shannon, H.S., Sullivan, T., Norman, R. and Wells, R.P., 1996, Disability resulting from occupational low back pain part II: What do we know about secondary prevention? A review of the scientific evidence on prevention after disability begins. *Spine*, **21**, pp. 2918-2929.

Frank, J.W. and Guzman, J., 1999, *Facilitation of RTW after a soft tissue injury: synthesizing the evidence and experience*: A HEALNet report on the findings of the Work-Ready project. HEALNet.

Gice, J.H. and Tompkins, K., 1989, RTW in a hospital setting. *Journal of Business and Psychology,* **4**, pp. 237-243.

Habeck, R.V., Leahy, M.J., Hunt, H.A., Chan, F. and Welch, E.M., 1991, Employer factors related to workers' compensation claims and disability management. *Rehab Counselling Bull*, **34**, pp. 210-226.

Himmelstein, J.S., Feuerstein, M., Stanek, E.J.I., Koyamatsu, K., Pransky, G.S., Morgan, W. and Anderson, K.O., 1995, Work-related upper-extremity disorders and work disability: clinical and psychosocial presentation. *Journal of Occupational and Environmental Medicine*, **35**, pp. 1278-1286.

Hirschfeld, R.M., Keller, M.B., Panico, S., Arons, B.S., Barlow, D., Davidoff, F., Endicott, J., Froom, J., Goldstein, M., Gorman, J., Guthrie, D., Marek, R.G., Maurer, T., Meyer, R., Phillips, K., Ross, J., Schwenk, T.L., Shaftstein, S.S., Thase, M. and Wyatt, R.J., 1997, The National Depressive and Manic-Depressive Association consensus statement on the undertreatment of depression. *Journal of the American Medical Association*, **277**, pp. 333-340.

Hlatky, M., Haney, T., Barefoot, J., Califf, R., Mark, D., Pryor, D. and Williams, R.B., 1986, Medical, psychological and social correlates of work disability among men with coronary artery disease. *American Journal of Cardiology*, **58**, pp. 911-915.

Hogelund, J., 2000, *Bringing the sick back to work: Labor market reintegration of the long-term sicklisted in the Netherlands and Denmark.* 1-258. Danish national Institute of Social Research. Roskilede University, Copenhagen. Ref Type: Thesis/Dissertation.

Hogg-Johnson, S. and Cole, D.C., 1998, *Early prognostic factors for duration on benefits among workers with compensated occupational soft tissue injuries.* (Toronto: Institute for Work & Health).

Hogg-Johnson, S., Frank, J.W., and Rael, E.G.S., 1994, *Prognostic risk factor models for low back pain: why they have failed and a new hypothesis.* Toronto: OWCI.

Hunt, H.A., Habeck, R.V., VanTol, B. and Scully, S.M., 1993, *Disability prevention among Michigan employers, 1988-1993.* Michigan: W.E. Upjohn Institute for Employment Research.

Hyatt, D., 1996, Work disincentives of workers' compensation permanent partial disability benefits: evidence for Canada. *Canadian Journal of Economics*, **29**, pp. 289-308.

Ison, T.G., 1986, The therapeutic significance of compensation structures. *The Canadian Bar Review,* **64**, pp. 605-637.

Jamison, R.N., Parris, W.C. and Maxson, W.S., 1987, Psychological factors influencing recovery from outpatient surgery. *Behaviour Research and Therapy,* **25**, pp. 31-37.

Johnson, W.G., Butler, R.J. and Baldwin, M., 1995, First spells of work absences among Ontario workers, In: Thomason T, Chaykowski RP, editors. *Research in Canadian workers' compensation.* (Kingston: IRC Press) pp. 73-84.

Joseph, J., Breslin, C. and Skinner, H., 1999, Critical perspectives on the transtheoretical model and stages of change. In: Tucker JA, Donovan DM, Marlatt GA, editors. *Changing Addictive Behaviors – Bridging Clinical and Public Health Strategies.* (New York: Guilford Press) pp. 161-190.

Kerns, R.D., Rosenberg, R., Jamison, R.N., Caudill, M.A. and Haythornthwaite, J., 1997, Readiness to adopt a self-management approach to chronic pain: the Pain Stages of Change Questionnaire (PSOCQ). *Pain*, **72**, pp. 227-234.

Kornblum, G.O., 1979, The role of the life, health and accident insurer's medical director in extra-contract claims litigation. *Transactions of the Association of Life Insurance Medical Directors of America*, **62**, p.61.

Kralj, B., 1995, Experience rating of workers' compensation insurance premiums and the duration of workplace injuries. In: Thomason T., Chaykowski R, (editors) *Research in Canadian Workers Compensation*. (Kingston: IRC Press) pp. 106-121.

Krause, N., Dasinger, L.K., Deegan, L.J., Brand, R.J. and Rudolph, L., 2001a, Psychosocial job factors and RTW after low back injury: A disability phase-specific analysis. *American Journal of Industrial Medicine*, **40** (4): pp. 464-484.

Krause, N., Dasinger, L.K. and Neuhauser, F., 1998a, Modified work and RTW: a review of the literature. *Journal of Occupational Rehabilitation*, **8**, pp. 113-139.

Krause, N., Frank, J.W., Sullivan, T.J., Dasinger, L.K. and Sinclair, S.J., 2001b, Determinants of RTW and Duration of Disability after Work-Related Injury and Illness: Challenges for Future Research, **40** (4): pp. 464-484.

Krause, N., Lynch, J., Kaplan, G.A., Cohen, R.D., Goldberg, D.E. and Salonen, J.T., 1997, Predictors of disability retirement. *Scandinavian Journal of Work Environment and Health*, **23**, pp. 403-413.

Krause, N., Ragland, D.R., Fisher, J.M. and Syme, S.L., 1998b, Psychosocial job factors, physical workload and incidence of work-related spinal injury: a 5-year prospective study of urban transit operators. *Spine*, **23**, pp. 2507-2516.

Krause, N. and Ragland, D.,R., 1994, Occupational disability due to low back pain: a new interdisciplinary classification based on a phase model of disability. *Spine*, **19**, pp. 1011-1020.

Lanoie, P., 1992, The impact of occupational safety and health regulation on the risk of workplace accidents: Quebec, 1983-87. *Journal of Human Resources*, **27**, pp. 643-660.

Linton, S.J., 2001, A review of psychological risk factors in back and neck pain. *Spine*, **25**, pp. 1148-1156.

Lippel, K., 1999, Therapeutic and anti-therapeutic consequences of workers' compensation. *International Journal of Law and Psychiatry*, **22**, pp. 521-546.

Loisel, P., Abenhaim, L., Durand, P., Esdaile, J.M., Suissa, S., Gosselin, L., Simard, R., Turcotte, J. and Lemaire, J., 1997, A population-based, randomized clinical trial on back pain management. *Spine*, **22**, pp. 2911-2918.

Loisel, P., Durand, P., Abenhaim, L., Gosselin, L., Simard, R., Turcotte, J. and Esdaile, J.M., 1994, Management of occupational back pain: the Sherbrooke model. Results of a pilot and feasibility study. *Occupational and Environmental Medicine*, **51**, pp. 597-602.

MacKenzie, E.J., Morris, J.A., Jurkovich, G.J., Yasui, Y., Cushing, B.M., Burgess, A.R., Andrew, R., DeLateur, B.J., McAndrew, M.P., Mark, P. and Swiontkowski, M.D., 1998, RTW Following Injury: The Role of Economic, Social and Job-Related Factors. *American Journal of Public Health*, **88**, pp. 1630-1637.

Maeland, J.G. and Havik, O.E., 1987, Psychological predictors for RTW after a myocardial infarction. *Journal of Psychosomatic Research*, **31**, pp. 471-481.

McIntosh, G., Frank, J., Hogg-Johnson, S., Bombardier, C. and Hall, H., 2000, Prognostic factors for time receiving workers' compensation benefits in a cohort of patients with low back pain. *Spine*, **25**, pp.147-157.

Mondloch, M.V., Cole, D.C. and Frank, J.W., 1999, Working Paper #72. *Does how you do depend how you think you'll do? A structured review of the evidence for a relation between patients' recovery expectations and outcomes*. (Toronto:

Institute for Work & Health), pp. 1-27.

Nachemson, A.L., 1983, Work for all: for those with low back pain as well. *Clinical Orthopaedics and Related Research*, **179**, pp. 77-85.

National Institute for Occupational Safety and Health. Musculoskeletal disorders and workplace factors, 1997. *A critical review of epidemiologic evidence for work-related musculoskeletal disorders of the neck, upper extremity, and low back.* (Baltimore: U.S. Department of Health and Human Services).

Norman, G., Fava, J.L., Levesque, D.A., Redding, C.A., Johnson, S., Evers, K. *et al.*, 1997, An inventory for measuring confidence to manage stress. *Annals of Behavioral Medicine*, **19**.

Papageorgiou, A.C., Macfarlane, G.J., Thomas, E., Croft, P.R., Jayson, M.I.V. and Silman, A.J., 1997, Psychosocial factors in the workplace: do they predict new episodes of low back pain? Evidence from the South Manchester Back Pain Study. *Spine*, **22**, pp.1137-1142.

Petrie, K., Weinman, J. and Sharpe, N., 1996, Role of patients' view of their illness in predicting RTW and functioning after myocardial infarction: longitudinal study. *British Medical Journal*, **312**, pp. 1191-1194.

Polatin, P.B., 1991, Predictors of low back pain disability. In: White AH, Anderson RT, editors. *Conservative Care of Back Pain.* (Baltimore: Williams & Wilkins) pp. 265-273.

Pransky, G., Benjamin, K., Hill-Fotouhi, C., Himmelstein, J., Fletcher, K.E., Katz, J.N. and Johnson, W.G., 2000, Outcomes in work-related upper extremity and low back injuries: Results of a retrospective study. *American Journal of Industrial Medicine*, **37**, pp. 400-409.

Pransky, G., Benjamin, K., Himmelstein, J., Mundt, K., Morgan, W., Feuerstein, M., Koyamatsu, K. and Hill-Fotouhi, C., 1999, Work-related upper-extremity disorders: prospective evaluation of clinical and functional outcomes. *Journal of Occupational and Environmental Medicine*, **41**.

Prochaska, J.O., Diclemente, C.C. and Norcross, J.C., 1992, In search of how people change. Applications to addictive behaviors. *American Psychologist*, **47**, pp. 1102-1114.

Prochaska, J.O., Velicer, W.F., Diclemente, C.C. and Fava, J., 1988, Measuring processes of change: applications to the cessation of smoking. *Journal of Consulting & Clinical Psychology*, **56**, pp. 520-528.

Prochaska, J.O., Velicer,W.F., Rossi, J.S., Goldstein, M.G., Marcus, B.H. and Rakowski, W., *et al.*, 1994, Stages of change and decisional balance for 12 problem behaviors. *Health Psychology*, **13**, pp. 39-46.

Ranney, D., 1997, *Chronic musculoskeletal injuries in the workplace.* (Philadelphia: W.B. Saunders), pp. 1-336.

Ruiz, Ba, Dibble, S.L., Gilliss, C.L. and Gortner, S.R., 1992, Predictors of general activity 8 weeks after cardiac surgery. *Applied Nursing Research*, **5**, pp. 59-65.

Sandstrom, J. and Esbjornsson, E., 1986, RTW after rehabilitation. The significance of the patient's own prediction. *Scandinavian Journal of Rehabilitation Medicine*, **18**, pp. 29-33.

Schade, B., Semmer, N., Main, C.J., Hora, J. and Boos, N., 1999, The impact of clinical, morphological, psychosocial and work-related factors on the outcome of lumbar discectomy. *Pain*, **80**, pp. 239-249.

Shanfield, S.B., 1990, RTW after an acute myocardial infarction: A review. Heart

& Lung. *The Journal of Critical Care*, **19**, pp. 109-117.

Sinclair, S., Hogg-Johnson, S., Erdeljan, S., and Mondloch, M., 1997, The effectiveness of an early active intervention program for workers with soft tissue injuries: The Early Claimant Cohort study. *Spine*, **22**, pp. 2919-2931.

Smith, J.M., Tarasuk, V., Shannon, H. and Ferrier, S., 1998, *ECC Prognosis Modelling Group. Prognosis of musculoskeletal disorders: effects of legitimacy and job vulnerability.*, IWH Working Paper #67 ed. (Toronto: Institute for Work & Health), pp. 1-15.

Smith, J., Tarasuk, V., Ferrier, S. and Shannon, H., 1996, Relationship between workers' reports of problems and legitimacy and vulnerability in the workplace a duration of benefits for lost-time musculoskeletal injuries. *American Journal of Epidemiology*, **143**, S17, Ref Type: Abstract.

Spitzer, W.O., LeBlanc, F.E., Dupuis, M., Abenhaim, L., Belanger, A.Y., Bloch. R., *et al.*, 1987, Scientific approach to the assessment and management of activity-related spinal disorders: a monograph for clinicians. Report of the Quebec task force on spinal disorders. *Spine*, **12**, pp. s4-s55.

Tarasuk, V. and Eakin, J.M., 1994, "Back problems are for life": perceived vulnerability and its implications for chronic disability. *Journal of Occupational Rehabilitation*, **4**, pp. 55-64.

Tarasuk, V. and Eakin, J.M., 1995, The problem of legitimacy in the experience of work-related back injury. *Qualitative Health Research*, **5**, pp. 204-221.

Thomason, T. and Pozzebon, S., 1995, The effect of workers's compensation benefits on claims incidence in Canada. In: Thomason T., Chaykowski R, editors. *Research in Canadian Workers' Compensation*. (Kingston: IRC Press) pp. 53-69.

van der Weide, W.E., Verbeek, J.H.A.M., Salle, H.J.A. and van Dijk, F.J.H., 1999, Prognostic factors for chronic disability from acute low-back pain in occupational health care. *Scandinavian Journal of Work, Environment and Health*, **25**, pp. 50-56.

Velicer, W.F., Norman, G.J., Fava, J.L. and Prochaska, J.O., 1999, Testing 40 predictions from the transtheoretical model. *Addictive Behaviors*, **24**, pp. 455-469.

Volinn, E., Van Koevering, D. and Loeser, J.D., 1991, Back sprain in industry: the role of socioeconomic factors in chronicity. *Spine*, **16**, pp. 542-548.

Waddell, G., Newton, M., Henderson, I., Somerville, D. and Main, C.J., 1993, A fear-avoidance beliefs questionnaire (FABQ) and the role of fear-avoidance beliefs in chronic low back pain and disability. *Pain*, **52**, pp. 157-168.

Waehrer, G. M., Miller, R., Ruser, J., and Leigh, J.P., 1998, *Restricted work, workers' compensation, and days away from work*. Unpublished Work.

Wilkinson, W.E., 1994, Therapeutic jurisprudence and workers' compensation. *Arizona Attorney*, **30**, pp. 28-33.

Yassi, A., Tate, R., Cooper, J.E., Snow, C., Vallentyne. S. and Khokhar, J.B., 1995, Early intervention for back-injured nurses at a large Canadian tertiary care hospital: an evaluation of the effectiveness and cost benefits of a two-year pilot project. *Occupational Medicine*, **45**, pp. 209-214.

Yelin, E.H., 1986, The Myth of Malingering: Why Individuals Withdraw from Work in the Presence of Illness. *Milbank Memorial Fund Quarterly/Health and Society*, **64**, pp. 622-647.

CHAPTER 3

Worker Accommodation, Clinical Intervention and Return to Work

Patrick Loisel and Marie-José Durand

"A little labor, much health" (Proverb)

3.1 INTRODUCTION

Evidence-based management of acute back pain has recently been promoted in guidelines produced by several task forces from a number of industrialized countries (Group CSA, 1994; Dutch Society of Occupational Physicians, 1999; AHCPR, 1994). All emphasize the same key points: rule out red flags (symptoms and clinical signs indicating a possible severe disease); avoid unnecessary imaging tests and treatments; recommend a quick return to normal activity including work. Subacute back pain is the clinical phase, as defined by Frank, that extends from four to twelve weeks of absence from activity and/or work (Frank *et al*, 1996a,b). At this time, if red flags have been ruled out, the narrowly-defined back pain problem becomes a much broader disability problem and should be considered as such (Fordyce, 1994). This is not to say that suffering should be ignored, but that it has to be addressed according to the person-environment model (Bronfenbrenner, 1979; Dobreen, 1994; World Health Organization, 2001; Parker *et al.*, 1989) rather than a pure biomedical model or even a bio-psychosocial model (Waddell *et al.*, 1984). Unfortunately, recent studies have shown the persistence of this prevailing biomedical model and its inability to prevent prolonged disability (Carey *et al.*, 2000; Cherkin *et al.*, 1996; Elam *et al.*, 1995) in the minority of cases that are responsible for most of the costs related to back pain (Spitzer *et al.*, 1987; van Tulder *et al.*, 1995) and that also have the worst health outcomes.

In this chapter, after a brief review of the current evidence concerning work disability prevention, we will emphasize the importance of situating work rehabilitation in the workplace, and present a novel intervention program we have developed and implemented to make it a reality.

3.2 RECENT EVIDENCE ON THE MANAGEMENT OF SUBACUTE BACK PAIN CASES LEADING TO OCCUPATIONAL DISABILITY

Studies over the past decade have brought new explanations and solutions for these subacute cases, issuing from epidemiological data on the causes of disability, and directly from intervention studies. Causes of prolonged disability due to back pain do not relate only on the back itself, but also to the whole person having back pain and to the social environment of the person, including the workplace. Also, studies over the past decade have shown that back pain management models that are disability-centered, rather than disease-centered, are more successful in terms of return to work (RTW) and well being. Conversely, during the acute or chronic stages, no single treatment modality was shown to be successful or more successful than placebo for achieving RTW (Frank *et al.*, 1996a,b; Waddell and Burton, 2000). Actually, as we will show, work disability is a complex issue influenced by multiple determining factors that go well beyond the pain in the back, and which originate from multiple messages and interactions among the worker as a physical, mental and social person, the workplace, the compensation system and also paradoxically the healthcare system (Figure 3.1).

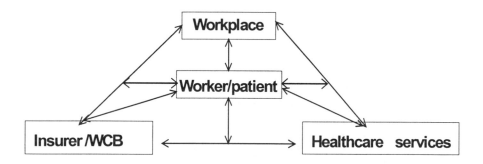

Figure 3.1 Interactions among stakeholders in the disability problem. The workers' disability is influenced by the stakeholders' actions and attitudes and by interactions occurring among stakeholders.

3.2.1 Causes of Work Disability

Some worker-related factors have been found consistently in several studies to be high-risk factors for prolonged disability: past history of work absence for back pain (Infante-Rivard and Lortie, 1996; Rossignol *et al.*, 1992), older age (Goertz, 1990; Infante-Rivard and Lortie, 1996), clinical severity (Abenhaim *et al.*, 1995), pain radiating below the knee (Selim *et al.*, 1998; van der Weide *et al.*, 1999). Also, psychological factors may be linked to prolonged work absenteeism but it remains unclear if these factors precede or follow the onset of back disability, with some evidence supporting each mechanism (Dionne, 1999; Nachemson, 1999; Truchon and Fillion, 2000). Interestingly, no factor clearly related to a specific anatomic or physiologic disorder or dysfunction has been related to prolonged disability. Disc degeneration, which was for years thought to be a disease, has now been repeatedly shown to be largely a normal aging process, often beginning early in the life. Many studies have shown the high prevalence (46% to 93%) of radiologically demonstrated disc degeneration at middle age and the poor relation between disc degeneration and symptoms (Boden *et al.*, 1996; Boos *et al.*, 2000; Jarvik and Deyo, 2000). Pain radiating below the knee, which was clearly related to prolonged disability in several studies, was not identified as clear radicular pain in most of them (Cherkin *et al.*, 1996; Infante-Rivard and Lortie, 1996; Loisel *et al.*, 2002; van der Weide *et al.*, 1999). The persistent pain perception itself may not be mainly due to peripheral nociceptive impulses at this phase. Pain is a complex phenomenon which is characterized by an unpleasant perception, always with sensorial and emotional components: it was defined by the International Association for the Study of Pain (Fordyce, 1994) as "an unpleasant sensory and emotional experience associated with actual or potential tissue damage, or described in terms of such damage". It may persist even if the disorder responsible for the initial pain has healed, due to several mechanisms, including a lowered threshold of interconnecting spinal neural cells, depending not only on peripheral impulses but also on central impulses from past and present cognition and emotions (Coderre *et al.*, 1993). This further level of neuronal involvement in pain perception was offered as a hypothesis by Melzack in 1965 (Melzack and Wall, 1965) in his "gate control theory". Since then, it has been confirmed and reinforced by numerous experimental studies (Coderre *et al.*, 1993). Mechanisms of chronic pain, also named centralization of pain, may include allodynia (enhancement of pain by reduction of the pain threshold), hyperalgesia (increased sensitivity to peripheral pain stimuli), persistent pain and referred pain. These biological phenomena explain how a high level of pain may be perceived by some patients even though the initial inducing lesion is healed. This can lead (through fear avoidance) to the unnecessary reduction of motion and activity, even though movement and function are not only deleterious but have in fact been shown to be helpful in enhancing healing (Fordyce, 1994).

Recently, special attention has been paid to the fear-avoidance behavior associated with the development of chronic musculoskeletal pain (Vlaeyen and Linton, 2000). Fear-avoidance, which refers to the avoidance of movements or activities based on the fear of doing them, has been put forth as a central mechanism in the development of chronic back problems. In particular, fear avoidance is thought to play an instrumental role in the deconditioning syndrome. Although some authors have called this phenomenon an irrational fear or phobia (Kori *et al.*, 1990), Vlaeyen has proposed a "fear avoidance model" (Vlaeyen *et al.*, 1995) based on previous work (Lethem *et al.*, 1983; Philips, 1987; Waddell *et al.*, 1993). This model suggests that the interpretation of the peripheral pain signal (catastrophizing versus no fear as extremes of the spectrum) may lead to disuse, depression and disability or, at the opposite end of the spectrum, to quick recovery. This is in accordance with the above-mentioned biological pain mechanisms. Fear seems to stay at the crossroads of the disability process.

Moving beyond worker factors, we will now briefly review factors not directly related to the worker. When the worker is absent from work, his or her natural environment is modified because relations with the workplace are deeply affected, and unusual relations with the healthcare system and with a disability compensation system occur.

Work demands have been shown to be responsible for back pain occurrence in many studies (Frank *et al.,* 1996a,b). However, until recently few epidemiological studies have demonstrated a direct link between physical work demands and work absenteeism (Dionne, 1999; Nachemson, 1999). Conversely, several studies have demonstrated the influence of the so called "psychosocial factors" related to the workplace, or to the interaction between the workplace and the worker, on work absenteeism for health reasons. Bigos (Bigos *et al.,* 1992) has shown in a large prospective study conducted in the Boeing industry that satisfaction with work at baseline influenced future absence from back pain. Recently, van der Weide (van der Weide *et al.,*1999) has found that perceived high work tempo and work quantity, and problems in relations with colleagues, predicted prolonged absence from work. Kerr has shown that perception that the "job is hard" predicted back pain reporting on the job (Kerr, 1997). Also, work organizational factors were recently demonstrated to influence RTW (Baril and Berthelette, 2000; Frank *et al.,* 1996a,b; Friesen *et al.,* 1999). In exploratory research, Baril and Berthelette (2000) found that the following structural characteristics of workplaces influenced the way early RTW measures were implemented: type and magnitude of allocated resources, economic sector, workplace size and financial health, unionization, characteristics of health and safety committees and presence of preventive programs. In addition, attitudes and values of health and safety personnel and attitudes of co-workers were of great influence on RTW (Baril and Berthelette, 2000). These results were supported by another study carried out in three provinces of Canada and in various industrial sectors (Group WRC, 1999). Key conditions found to facilitate successful management of workers having musculoskeletal disorders included effective communication, collaboration and trust among the various stakeholders involved, such as attending physicians and Workers' Compensation Board (WCB) case-workers. Employers' willingness to get their workers back to work seems also to influence disability duration and costs (Orme *et al.,* 2000; Shrey, 1996).

Several studies indicate that recent scientific evidence on back pain management is not yet being applied by health care providers (Battie *et al.,* 1994; Elam *et al.,*1995; Rainville *et al.,* 2000; van Tulder *et al.,* 1997). In particular, unnecessary imaging tests and prolonged unnecessary treatment appear to have a labelling effect, with deleterious consequences on the worker (Nachemson, 1999).

When a worker is absent from work due to back pain as a work-related health problem and is compensated for that, relations with the healthcare system and providers are more complex than normal patient-provider interactions (Figure 3.1). Providers have to formally estimate the time of RTW and recommend work restrictions. For this reason, most insurance or WCB regulations formally require information from both the worker and the providers. A lack of communication between healthcare providers and the workplace may also have a negative effect on RTW (Friesen *et al.,* 1999). Also, some authors claim that the level of the compensation for work absence has a direct influence on the duration of absence for back pain (Nachemson, 1999). Moreover, the compensation system sometimes generates adversarial legal actions that have been shown to have a deleterious effect on disability (Baril *et al.,* 1994; Butterfield *et al.,* 1998; Nachemson, 1999).

3.2.2 Intervention Studies of Disability Prevention at the Subacute Phase

During the past few years, several models of disability prevention for low back pain have been developed and tested. The most promising intervention studies have implemented a disability prevention model during the subacute phase of the disease.

Lindstrom (Lindstrom *et al.*, 1992) has shown, in a randomized clinical trial, that workers with subacute back pain treated with an intervention of graded exercises closely linked to the worker's job in the workplace had significant less time on sick leave, and less recurrences, than those treated with usual health care. Indahl (Indahl *et al.*, 1995) has shown in a randomized clinical trial that giving subacute back pain patients extensive and repeated reassuring explanations on their back pain could cut by half the duration of absence from work. Yassi (Yassi *et al.*, 1995) has shown the effectiveness and cost effectiveness of an early intervention (including work rehabilitation and job modifications) in a population of nurses in a large university hospital in Manitoba. Loisel *et al.* (1997) has shown in a population-based randomized clinical trial that a model of subacute back pain management (Sherbrooke model), linking a rehabilitation intervention and a workplace intervention including job modifications, sped up the return to regular work by 2.4-fold ($p=0.01$); the most important effect came from the workplace intervention that sped up RTW by 1.9-fold ($p<0.01$). Also, functional status was improved and pain level was reduced by the interventions. The interventions included in this model of management had as their principal aim that early identification in the participating workplaces of the workers at risk of prolonged disablement (at four weeks of absence from regular work). The model also helped them to RTW using a work rehabilitation process, graded to match the evolution of the worker's progress in healing (Loisel *et al.*, 1994). In other words, interventions were delivered when needed by a multidisciplinary team, via a rehabilitation process named "therapeutic RTW" (TRW). This process was centered in the workplace and return to regular work was initiated at the earliest possible time (Durand and Loisel, 2000). Occupational interventions included visits to an occupational medicine physician, as well as a participatory ergonomics intervention involving an ergonomist, the injured worker, his or her supervisor and management and union representatives. Job modifications were recommended to the employer, who was at liberty to implement them or not. The clinical/rehabilitation intervention consisted of a clinical examination by a back pain medical specialist, participation in a "back school" after eight weeks of absence from regular work and, if necessary, the TRW process. These progressively more intensive interventions were gradually offered following a defined schedule, if regular work was not already resumed.

Lindstrom's (Lindstrom *et al.*, 1992) and Loisel's (Loisel *et al.*, 2001b) trials were also shown to be cost beneficial. Thus there is now accumulating evidence that the subacute phase could be the "golden hour of opportunity" for preventing low back pain disability. However, the ideal intervention recipe is not clear, as the above-mentioned trials were using different intervention packages. However, some common principles may be derived from these studies. Successful interventions all involved professional staff independent from the employer or the insurer, who reassured the patients after careful examination, and dissuaded them from extensive investigations and treatments. If these investigations were done (prescribed or not by the study staff), explanations of the results were given in a reassuring way. In all studies, large amounts of time were spent with the patients to make sure that they understood the reassuring messages. In Indahl's study (Indahl *et al.*, 1995), this was the only formal intervention, but it was not clear if the back pain of the study cases was work-related, as it was in the other studies. This could explain the apparently large effect demonstrated in spite of the absence of a workplace intervention, which was shown to be effective in other studies. Also, even if evidence of effectiveness for RTW of programs using partial or progressive duties is limited, Krause (Krause *et al.*, 1998) has shown in a recent review that their efficacy is likely. What can be concluded to date from these studies is that bringing to the patient a coherent, organized evaluation and case management, delivered by an independent team working in close coordination with the employer and the insurer, is more effective for return-to-work and for quality-of-life than usual management. Reassuring explanations and links with the workplace seem to be key issues.

3.3   PROPOSAL FOR A REFINED WORK DISABILITY PREVENTION MANAGEMENT PROGRAM IN THE SUBACUTE OR EARLY CHRONIC PHASE

From this evidence we designed a program that would directly address the possible causes of disability for every worker at the subacute or early chronic phase, and that could be implemented in the Quebec setting. We were able to benefit from previous Sherbrooke model experience (Loisel *et al.*, 1997) and from the support of the public health division of the district of Montérégie (a large area located south of Montreal). In order to design a methodologically strong study, we set up a workgroup made of researchers and clinicians in the field of occupational back pain disability (occupational physicians, back pain specialist, occupational therapist, ergonomist and statistician),[1] some of whom had participated in the Sherbrooke trial. The workgroup agreed on the following points.

   a)   The PREVICAP (PREVention of work handICAP) program should have two main steps (Figure 3.2):

   •   a "disability diagnosis" step to identify, in each worker's case, the precise causes of disability (physical, psychosocial, occupational and administrative);
   •   a progressive RTW process: TRW, as it had been designed for the Sherbrooke trial (Durand *et al.*, 1998; Loisel *et al.*, 1994), associated with an ergonomic intervention at the workplace, for the "worker's regular job" (Loisel *et al.*, 2001a).

   b)   The program should be hospital-based (in the Quebec public hospital system), initiated at the subacute phase of back pain and delivered as specialty care rather primary care.

c) The program should involve an interdisciplinary team to be able to address the various medical and non-medical causes of disability.
d) The program should interact closely with the worker's attending physician, the workplace and the Quebec WCB (CSST).

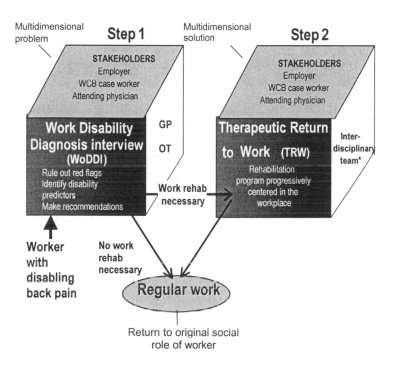

Interdisciplinary team: GP, OT, ergonomist, kinesiologist, psychologist, musculoskeletal specialist, team coordinator.

Figure 3.2 Schema of the PREVICAP program.

The work disability diagnosis (WoDDI) (Durand *et al.*, 2002) step was not described in the Sherbrooke model, but, in practice, a WoDDI process was informally carried out by the multidisciplinary clinical team involved in the trial, in addressing the problems impeding RTW for each case. The WoDDI step is founded on the necessity of identifying in each worker's case the prolonged disability predictors that are rarely the original disease itself (back pain) but most often all other predictors, according to the work disability paradigm discussed above and elsewhere (Loisel *et al.,* 2001c).

To ensure that the WoDDI process could systematically be carried out at an early stage, potential disability prognosis factors, based on scientific evidence and clinical experience, were identified by the workgroup and classified as physical, psychosocial, occupational and administrative factors. Questions about all these possible factors have been developed and collated into an interview form used at the first encounter with the worker. Grossly, it includes the usual medical questionnaire and physical examination regarding: back pain, work history, work demands, job difficulty, life habits, family and social history, perceptions concerning RTW and an assessment of the worker's financial situation. Also, questions designed to identify possible clinical "red flags" were included. This WoDDI tool is jointly administered by a work rehabilitation physician for its medical and psychosocial part, and by an occupational therapist (OT) for its occupational part. Then, the physician and the OT meet to share their findings and design a treatment plan that is recommended to the worker, the attending physician and the WCB case-worker. In some cases, barriers to RTW appear to be minimal: reassurance of the worker and other stakeholders appears sufficient to allow a quick return to regular work (RTRW) at minimal cost. Otherwise, the treatment plan is the TRW.

The second step, tailored to the worker's precise needs, identified in the WoDDI step, is the TRW, extending from the beginning of the work rehabilitation intervention to the full return to regular or permanently modified work. TRW resembles a customized management plan more than a treatment, because it adds to treatment modalities actions for the other stakeholders (besides the patient) to take, and includes interventions in the workplace. It corresponds to a comprehensive rehabilitation process, addressing directly the overall disability problem rather than the disease, with the explicit purpose of restoring the patient's social role as worker (Loisel *et al.,* 2001a).

One fundamental characteristic of the TRW is that it is situated in the workplace instead of in a clinical setting, as is more usual in work rehabilitation interventions (Durand *et al.,* 1998). The workplace is not considered as a potentially harmful place, but rather as a rehabilitation setting, insofar as its potentially harmful exposures can be appropriately controlled by the rehabilitation team, in close collaboration with the employer. As this intervention involves, as a principal component, a return of the disabled worker to the job, it does not depend only on the worker and on the clinical team. The employer (and, if applicable, the union), the worker's attending physician and the WCB case-worker have to be part of the enlarged RTW team. The usual bipartite healthcare encounter, HCP-patient, becomes a five-player team: patient/worker, work rehabilitation interdisciplinary team, employer, attending physician and WCB case-worker. All have to agree to make efforts to attain a common aim: RTW. These collaborative efforts may be demanding but they will be rewarded through engagement and empowerment of all parties whose joint efforts are necessary for successful RTW.

The worker's responsibility is to accept the difficult process of regaining physical and mental capacities and to go through fears and pain. In return, the worker not only receives the help of a rehabilitation team for clinical interventions, but also receives help inside the workplace to decrease regular job demands and to provide progressive, graded exposure to these regular job demands, according to his or her improving capacity. This approach is very different from the modified work (light duties) approach, which usually exposes the worker to job tasks different from those of the worker's regular work. These unusual job tasks may not represent progressive retraining to the regular job, and may also have a negative effect on the worker's perceptions of the usefulness of these tasks to improve his or her condition (Durand *et al.*, 1998).

The employer's responsibility is to accept a progressive return to regular work (instead of a straightforward return to full duties), an ergonomist's visit on the job, followed by recommended job modifications, their implementation, and an interaction with a clinical team – who determines week by week the worker's production capabilities. In return, the employer reintegrates a satisfied productive worker and often obtains ergonomic job improvements that may spin off to the benefit of other workers. In Quebec, due to WCB regulations, the worker may be paid directly by the employer in a progressive fashion as RTW progresses, instead of staying on full benefits: due to compensation regulations, this will allow the employer to save large amounts of money on his or her future WCB bill. Also, the worker has by law the right to return to his or her regular job any time up to two years following the work accident (one year for small employers with less than 20 workers). This job security legislation facilitates reintegration to the regular job if sufficient well being is achieved after injury, and/or appropriate job modifications are implemented. In other respects, the employer will be charged for all WCB expenses related to the work accident (including reserves for future costs) and occurring during the four years following it. With such regulations, the cost of prolonged disability may be considerable, making work rehabilitation an attractive option for the employer.

The worker's attending physician's responsibility is to consider the interdisciplinary team as a legitimate work rehabilitation "specialist" (this is presently unusual) and to reinforce the reassuring messages of the team to the worker. The physician has also to accept the whole disability management paradigm, with its corollary of treating the work disability rather than just the disease, and to have confidence in the team, thinking of them as safely conducting the return to the workplace as a therapeutic option. In return, the physician is helped with the difficulty of returning such challenging patients to full function and is relieved of the time-consuming and underpaid tasks of interacting with the employer and the WCB case-worker.

The WCB case-worker's responsibility is to accept a share in the leadership of the RTW process with the interdisciplinary team, to facilitate the dialogue with the employer and to avoid legal actions. We favor an early joint visit to the employer by the interdisciplinary team ergonomist (or OT) and the WCB case-worker, in such a way that, at the same time, the employer is reminded of the worker's rights and the possible costs linked to prolonged disability (by the case-worker), and is presented with the work rehabilitation plan to avoid disability and promote return to the regular job (by the interdisciplinary team). In return, the case-worker receives help with difficult cases and can often settle the case definitively, without a future burden of permanent disability, pension or vocational rehabilitation.

Thus, TRW is a win-win process. However, reassuring a fearful deconditioned worker having persistent pain, and getting together stakeholders who are not usually accustomed to speaking with one another, may appear as a challenge. Actually, it is a challenge that requires some special efforts, especially from the interdisciplinary team. One of the roles of the team is to ensure that at any time the RTW remains a common aim, shared by all stakeholders, and that the worker is empowered and consistently given the same unique message concerning his or her condition, progress and RTW. This requires the team to maintain a very close collaboration, confidence and trust among its members and to develop a communication network with other stakeholders, adapted to each case. This teamwork is very different from usual interactions between healthcare professionals, where everyone offers his or her individual opinion on the case and tells the worker his or her own view of the problem, resulting in confusion for the worker. This interdisciplinary teamwork comes close to the high-performance team work encountered in highly effective organizations. Katzenbach and Smith have described effective teams in the following way: "a team is a small number of people with complementary skills who are committed to a common purpose, performance goals, and approach for which they hold themselves mutually accountable" (Katzenbach and Smith, 1994). The professionals of the team bring their complementary skills and common values to the common purpose of RTW and they know that they share among themselves, and with the other stakeholders, part of the responsibility for success or failure of the RTW process.

TRW is both an interdisciplinary and inter-organizational process addressing directly all the factors impeding the RTW at the interfaces between the stakeholders in the process, including the worker. It applies the scientific evidence on disability predictors to the idiosyncratic situation of each worker's work disability. Also, the TRW process is very different from in-depth *physical capacity evaluations* in the clinical setting or extensive major ergonomic assessments in the workplace, as may be made to modify entire production lines for primary prevention of injury at work. TRW makes physical and mental capacity assessments but these are set within a continuous process where more job capability is gradually matched to more appropriate job demands. It brings the opportunity to demonstrate steady rehabilitation progress to the worker and to the employer. When a difficulty occurs during the RTW process, it should be immediately addressed by the team – adjusting posture, job demands, attitudes, beliefs, as all of them may be responsible for the fear of RTW and pain magnification. The disability predictors which apply to the specific worker are directly addressed in an integrated physical, cognitive, behavioral, humanistic way, using various modes of communication and education. At the end of the process, the worker is working full duties instead of being declared "fit to work" without knowing his or her actual capabilities and behavior in the real job environment.

Team members are professionals able to interact with the patient/worker's physical and mental condition in an active way, as well as with the workplace, bridging the usual gap between the clinical setting and the workplace. The worker's physical condition needs to be improved; postural habits in daily life and at the job need to be addressed; back care and coping skills for chronic pain must be taught; fears and depression are frequent issues; job design often needs improvement; symptoms, medication, beliefs are also issues to deal with. To address jointly all these issues we have brought together the following professionals: a general practitioner (with skills in musculoskeletal disorders and the rehabilitation process), an OT, a physical educator (kinesiologist), a psychologist (skilled in pain issues), an ergonomist and a team coordinator who ensures the smooth functioning of the team with respect to values, purpose and performance. A back pain specialist is available for appropriate counseling on the disease, and reassurance of the team and the worker, especially in case of symptom augmentation during the rehabilitation.

The whole team develops a consensus on graded treatment and RTW, on messages to deliver to the patient and on communication strategy with stakeholders. However, the team might be somewhat different in its composition from country to country and across practice settings. The important thing is that team members are able to work together, have a common purpose, are mutually accountable, meet minimally every week to review every ongoing case, and can put aside interpersonal and inter-professional barriers. They recognize that, as no specific intervention has been shown uniformly efficacious in reducing or preventing prolonged disability from back pain, only a full team will be able to treat the disability. Also, every week, the team will design the appropriate strategy to ensure the stakeholders will not react inappropriately to the particular complications in a way that could interfere with rehabilitation (pain augmentation, workplace issue, differing opinion on the disease from another consultant, unexpected claims management notice from the WCB, etc.) (Figure 3.3).

The right timing for this kind of intervention is subject to debate and there are presently few precise scientific grounds for rigid adherence to any one schedule. Frank (Frank *et al.*, 1996) has indicated that, following present evidence, the most appropriate time appears to be the subacute phase of back pain (one to three months on benefits) and we share this opinion. However, in our experience, boards and physicians are often reluctant to refer the patients at this early stage. WCBs may think that these interventions are costly and prefer to wait to be sure the case may not be resolved by "usual care". HCPs may make prolonged efforts to try to cure the injury biomedically and are infrequently aware of the broader disability problem. In our experience (excluding accrual for RCTs, which required early timing as a study inclusion criterion), the mean reference time was ten months on benefits. However, even with these very disabled patients, we have obtained a 60% stable RTW rate at one year follow-up. A larger experiment, including four rehabilitation centers in four distinct geographic areas of the province of Quebec, is presently underway.

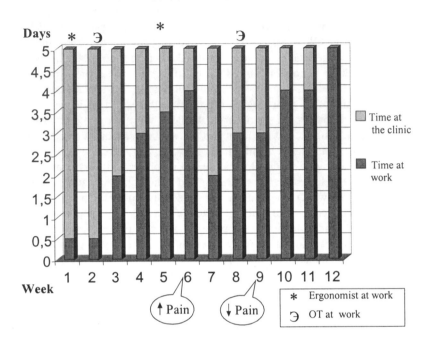

Figure 3.3 An example of progressive return to regular duties, adjusted to the current capacities of the worker and to the implementation of job modifications.

3.4 RESTORING THE WORKER'S ROLE

Recent evidence demonstrates that bringing to the patient a coherent, organized evaluation and case-management strategy, delivered by an independent team in close collaboration with the employer and the insurer, is more effective for RTW and restoring quality of life than usual management. Reassuring explanations and close links with the workplace are the key requisites. The bio-psychosocial and person-environment models must take the place of the traditional biomedical model, in that the evolving disability encompasses the disease. This is opposite to the usual biomedical management philosophy which consists mostly of brief encounters with the one healthcare provider at a time, and high-tech investigations, reinforcing in the patient's mind the thought that something is seriously wrong in his or her back, and he or she is probably at risk for worsening, especially if work is resumed (Elam *et al.,* 1995; van Tulder *et al.*, 1997).

The comprehensive disability prevention program described above takes into account present scientific evidence to bring empowerment to the patient/worker through progressive return to regular work. The process avoids evaluations and decisions taken far from the workplace context and reality. It ensures that the fearful worker is not subjected to contradictory explanations and decisions. Rather, it empowers the worker for restoring progressively his or her own capability of returning to a productive working life, in spite of some residual pain. Programs to help workers resume a valuable productive life may differ depending on the local social and political context, laws and regulations. The PREVICAP program was developed for application in the province of Quebec and different solutions may be found in other locations. However, in our opinion, the following basic principles should be applied in all settings: an interdisciplinary team approach, involving all stakeholders, and identifying and addressing each patient's disability determinants, with the primary purpose of achieving progressive return to regular work. Back pain should not deprive human beings of their working role. Graded and controlled return to regular work for back pain sufferers contributes to workplaces and workers' health.

3.5 NOTES

[1]Marie-José Durand, Lise Gosselin, Jacques Lemaire, Patrick Loisel, Serge Marquis, Alain Messier, Claude Prévost and Susan Stock.

3.6 KEY MESSAGES

- Work disability is influenced by multiple factors including interaction between stakeholders.

- The bio-psychosocial and person-environment models should replace the biomedical model in addressing the disability problem.

- A coherent, organized evaluation and case management, delivered by an independent interdisciplinary team in collaboration with the employer and the insurer, is more effective for RTW and restoring quality of life than "usual care".

- Reassuring explanations and links with the workplace are the key issues.

- Effective work rehabilitation programs should use an interdisciplinary and inter-organizational approach adapted to the regional social and political context.

3.7 ACKNOWLEDGEMENTS

This work was supported by grants from the Institut Robert-Sauvé de recherche en Santé et Sécurité du Travail (IRSST), Health Evidence Application and Linkage Network (HEALNet), Fonds de recherche en Santé du Québec (FRSQ) and the Régie Régionale de la Santé et des Services Sociaux de la Montérégie (RRSSSM).

3.8 REFERENCES

Abenhaim, L., Rossignol, M., Gobeille, D., Bonvalot, Y., Fines, P. and Scott S., 1995, The prognostic consequences in the making of the initial medical diagnosis of work-related back injuries. *Spine*, **20**, pp. 791-795.

Agency for Health Care Policy and Research. *Clinical practice guideline: Acute low back problems in adults*, Rockville, US Department of Health and Human Services, 1994.

Baril, R., and Berthelette, D., 2000, *Components and organizational determinants of workplace interventions designed to facilitate early RTW*, Montréal, Institut de recherche Robert-Sauvé en santé et sécurité du travail.

Baril, R., Martin, J.C., Lapointe, C. and Massicotte, P., 1994, *Étude exploratoire des processus de réinsertion sociale et professionnelle des travailleurs en réadaptation*, Montréal, Institut de Recherche en Santé et Sécurité du Travail du Québec,.

Battie, M.C., Cherkin, D.C., Dunn, R., Ciol, M.A. and Wheeler, K.J., 1994, Managing low back pain: attitudes and treatment preferences of physical therapists. *Physical Therapy*, **74**, pp. 219-226.

Bigos, S.J., Battie, M.C., Spengler, D.M., Fisher, D.M., Fordyce, W.E., Hansson, T., Nachemson, A.L. and Zeh, J., 1992, A longitudinal, prospective study of industrial back injury reporting. *Clinical Orthopaedics & Related Research*, pp. 21-34.

Boden, S.D., Riew, K.D., Yamaguchi, K., Branch, T.P., Schellinger D. and Wiesel, S.W., 1996, Orientation of the lumbar facet joints: association with degenerative disc disease. *Journal of Bone & Joint Surgery* – American Volume, **78**, pp. 403-411.

Boos, N., Semmer, N., Elfering, A., Schade, V., Gal, I., Zanetti, M., Kissling, R., Buchegger, N., Holder, J. and Main, C.J., 2000, Natural history of individuals with asymptomatic disc abnormalities in magnetic resonance imaging: predictors of low back pain-related medical consultation and work incapacity. *Spine*, **25**, pp. 1484-1492.

Bronfenbrenner, U., 1979, *The ecology of human development experiments by nature and design*. (Cambridge, MA: Harvard University Press).

Butterfield, P.G., Spencer, P.S., Redmond, N., Feldstein, A. and Perrin, N., 1998, Low back pain: predictors of absenteeism, residual symptoms, functional impairment, and medical costs in Oregon workers' compensation recipients. *American Journal of Industrial Medicine*, **34**, pp. 559-567.

Carey, T.S., Garrett, J.M. and Jackman, A.M., 2000, Beyond the good prognosis. Examination of an inception cohort of patients with chronic low back pain. *Spine*, **25**, pp. 115-20.

Cherkin, D.C., Deyo, R.A., Street, J.H. and Barlow, W., 1996, Predicting poor outcomes for back pain seen in primary care using patients' own criteria. *Spine*, **21**, pp. 2900-2907.

Coderre, T.J., Katz, J., Vaccarino, A.L. and Melzack, R., 1993, Contribution of central neuroplasticity to pathological pain: review of clinical and experimental evidence. *Pain*, **52**, pp. 259-285.

Dionne, C.E., 1999, *Low back pain, in Crombie IK, Epidemiology of pain*, (Seattle: IASP Press).

Dobreen, A.A., 1994, An ecological oriented conceptual model of vocational rehabilitation of people with acquired midcareer disabilities. *Rehabilitation Counseling Bulletin*, **37**, pp. 215-227.

Durand, M-J., Loisel, P. and Durand, P., 1998, Le Retour Thérapeutique au Travail comme une intervention de réadaptation centralisée dans le milieu de travail: description et fondements théoriques. *La revue canadienne d'ergothérapie*, **65**, pp. 72-80.

Durand, M-J. and Loisel, P., 2000, Therapeutic RTW: Rehabilitation in the workplace. *Work*, **17**, pp. 57-63.

Durand, M-J, Loisel, P., Hong, Q.N. and Charpentier, N., 2002, Helping clinicians in work disability prevention: the Work Disability Diagnosis Interview. *Journal of Occupational Rehabilitation*, accepted.

Dutch Society of Occupational Physicians., 1999, *Practice guidelines for low back pain*.

Elam, K.C., Cherkin, D.C. and Deyo, RA., 1995, How emergency physicians approach low back pain. choosing costly options. *Journal of Emergency Medicine*, **13**, pp. 143-150.

Fordyce, W.E., 1994, *Back pain in the workplace. Management of disability in non specific conditions* (Seattle: IASP Press).

Frank, J.W., Brooker, A.S., DeMaio, S.E., Kerr, M.S., Maetzel, A., Shannon, H.S., and Sullivan, T.J., 1996a, Disability resulting from occupational low back pain. Part II: What do we know about secondary prevention? A review of the scientific evidence on prevention after disability begins. *Spine*, **21**, pp. 2918-2929.

Frank, J.W., Kerr, M.S., Brooker, A.S., Norman, R.W., and Wells, R.P., 1996b, Disability resulting from occupational low back pain. Part I: What do we know about primary prevention? A review of the scientific evidence on prevention before disability begins. *Spine*, **21**, pp. 2908-2917.

Friesen, M.N., Yassi, A. and Cooper, J., 1999, Workready Manitoba: Stakeholder perspectives on return-to-work, *HEALNet Workready Research Group*.

Goertz, M.N., 1990, Prognostic indicators for acute low-back pain. *Spine*, **15**, pp. 1307-1310.

Group CSA., 1994, Back pain: *Report of a CSAG Committee on Back Pain*, London, HMSO.

Group WRC (Work Ready Collaborative), 1999, *Facilitation of RTW after a soft tissue injury: Synthesizing evidence and experience*, Report of Work Ready Phase 1 to HealNet.

Indahl, A., Velund, L. and Reikeraas, O., 1995, Good prognosis for low back pain when left untampered. A randomized clinical trial. *Spine*, **20**, pp. 473-477.

Infante-Rivard, C. and Lortie, M., 1996, Prognostic Factors For RTW After a First Compensated Episode of Back Pain. *Occupational & Environmental Medicine*, **53**, pp. 488-494.

Jarvik, J.G. and Deyo, R.A., 2000, Imaging of the lumbar intervertebral disc degeneration and aging, excluding disc herniations. *Radiology Clinics of North America*, **38**, pp. 1255-1266.

Katzenbach, J.R. and Smith, D.K., 1994, *The wisdom of teams. Creating the high-performance organization*. (New York: Harper Collins Publishers).

Kerr, M.S., 1997, A case-control study of biomechanical and psychosocial risk factors for low-back pain reported in an occupational setting, Doctoral Dissertation, *Graduate Department of Community Health*, Toronto: University of Toronto.

Kori, S.H., Miller, R.P. and Todd, D.D., 1990, Kinesophobia: A new view of chronic pain behavior. *Pain Management*, pp. 35-43.

Krause, N., Dasinger, L.K. and Neuhauser, F., 1998, Modified work and RTW: a review of the literature. *Journal of Occupational Rehabilitation*, **8**, pp. 113-139.

Lethem, J., Slade, P.D., Troup, J.D. and Bentley, G., 1983, Outline of a Fear-Avoidance Model of exaggerated pain perception – I. *Behaviour Research & Therapy*, **21**, pp. 401-408.

Lindstrom, I., Ohlund, C., Eek, C., Walin, L., Peterson, L.E., Fordyce, W.E. and Nachemson, A.L., 1992, The effect of graded activity on patients with subacute low back pain: a randomized prospective clinical study with an operant-conditioning behavioral approach. *Physical Therapy*, **72**, pp. 279-293.

Loisel P, Abenhaim L, Durand P., Esdaile, J.M., Suissa, S., Gosselin, L., Simard, R., Turcotte, J. and Lemaire, J., 1997, A population-based, randomized clinical trial on back pain management. *Spine*, **22**, pp. 2911-2918.

Loisel, P., Durand, M-J., Berthelette, D., *et al.*, 2001a, Disability prevention: the new paradigm of management of occupational back pain. *Disease Management & Health Outcomes*, **9**, pp. 351-360.

Loisel, P., Durand, P., Abenhaim, L., Gosselin, L., Simard, R., Turcotte, J. and Esdaile, J.J., 1994, Management of occupational back pain: the Sherbrooke model. Results of a pilot and feasibility study. *Occupational & Environmental Medicine*, **51**, pp. 597-602.

Loisel, P., Gosselin, L., Durand, P., Lemaire, J., Poitras, S. and Abenhaim, L., 2001b, Implementation of a participatory ergonomics program in the rehabilitation of workers suffering from subacute back pain. *Applied Ergonomics,* **32,** pp. 53-60.

Loisel, P., Lemaire, J., Poitras, S. *et al.*, 2001c, Cost-benefit and cost-effectiveness analysis of the Sherbrooke model of back pain management. *Occupational & Environmental Medicine*, Submitted.

Loisel, P., Vachon, B., Lemaire, J., Durand, M-J., Poitras, S., Stock, S. and Tremblay, C., 2002, Discriminative and predictive validity assessment of the Quebec Task Force Classification. *Spine*, **27**, pp. 851-857.

Melzack, R and Wall, P.D., 1965, Pain mechanisms: a new theory. *Science*, **150**, pp. 971-979.

Nachemson, A., 1999, Back pain: delimiting the problem in the next millennium. *International Journal of Law & Psychiatry*, **22**, pp. 473-490.

Orme, T.J., Covatta, C., Chappuis, J.L. and Wechsler, R.P., 2000, Employer attitude as a predictor of treatment and disability costs, *North American Spine Society 15th Annual Meeting, New Orleans*.

Parker, R.M., Szymanski, E.M., and Hanley-Maxwell, C., 1989, Ecological assessment in supported employment. *Journal of Applied Rehabilitation Counseling*, **20**, pp. 26-33.

Philips, H.C., 1987, Avoidance behavior and its role in sustaining chronic pain. *Behavioral Research Therapy*, **25**, pp. 273-279.

Rainville, J., Carlson, N., Polatin, P., Gatchel, R.J. and Indahl, A., 2000, Exploration of physicians' recommendations for activities in chronic low back pain. *Spine*, **25**, pp. 2210-2220.

Rossignol M., Suissa, S. and Abenhaim, L., 1992, The evolution of compensated occupational spinal inj,uries. A three-year follow-up study. *Spine*, **17**, pp. 1043-1047.

Selim, A.J., Ren, X.S., Fincke, G., Deyo, R.A., Rogers, W., Miller, D., Linzer, M. and Kazis, L., 1998, The importance of radiating leg pain in assessing health outcomes among patients with low back pain. Results from the Veterans Health Study. *Spine*, **23**, pp. 470-474.

Shrey, D.E., 1996, Disability management in industry: the new paradigm in injured worker rehabilitation. *Disability & Rehabilitation*, **18**, pp. 408-414.

Spitzer, W.O., LeBlanc, F.E. and Dupuis, M., 1987, Scientific approach to the assessment and management of activity-related spinal disorders. A monograph for clinicians. Report of the Quebec Task Force on Spinal Disorders. *Spine*, **12**, pp. S1-59.

Truchon, M. and Fillion, L., 2000, Biopsychosocial determinants of chronic disability and low-back pain: A review. *Journal of Occupational Rehabilitation*, **10**, pp.117 142.

van der Weide, W.E., Verbeek, J.H., Salle, H.J. and van Dijk, F.J., 1999, Prognostic factors for chronic disability from acute low-back pain in occupational health care. *Scandinavian Journal of Work, Environment & Health*, **25**, pp. 50-56.

van Tulder, M.W., Koes, B.W., Bouter, L.M. and Metsemakers, J.F., 1997, Management of chronic nonspecific low back pain in primary care: a descriptive study. *Spine*, **22**, pp. 76-82.

van Tulder, M.W., Koes, B.W. and Bouter, L.M., 1995, A cost-of-illness study of back pain in The Netherlands. *Pain*, **62**, pp. 233-240.

Vlaeyen, J.W., Kole-Snijders, A.M.J., Boeren, R.G.B. and van Eek, H., 1995, Fear of movement/(re)injury in chronic low back pain and its relation to behavorial performance. *Pain*, **62**, pp. 363-372.

Vlaeyen, J.W. and Linton, S.J., 2000, Fear-avoidance and its consequences in chronic musculoskeletal pain: a state of the art. *Pain*, **85**, pp. 317-332.

Waddell, G. and Burton, A.K., 2000, *Occupational health guidelines for the management of low back pain at work-evidence review*, London, Faculty of Occupational Medicine.

Waddell, G., Main, C.J., Morris, E.W., Di Paola, M. and Gray, I.C., 1984, Chronic low-back pain, psychological distress, and illness behavior. *Spine*, **9**, pp. 209-213.

Waddell, G., Newton, M., Henderson, I., Somerville, D. and Main, C.J., 1993, A Fear-Avoidance Beliefs Questionnaire (FABQ) and the role of fear-avoidance beliefs in chronic low back pain and disability. *Pain*, **52**, pp.157-168.

World Health Organization, 2001, International Classification of functioning, disability and health. *Geneva: World Health Organization*.

Yassi, A., Tate, R., Cooper, J.E., Snow, C., Vallentyne, S. and Khokhar, J.B., 1995, Early intervention for back-injured nurses at a large Canadian tertiary care hospital: an evaluation of the effectiveness and cost benefits of a two-year pilot project. *Occupational Medicine*, **45**, pp. 209-214.

CHAPTER 4

Injury Prevention and Return to Work: Breaking Down the Solitudes

Annalee Yassi, Aleck Ostry and Jerry Spiegel

4.1 INTRODUCTION

Work injuries arise from complex interactions between individual workers and their work environments (Verbrugge and Jette, 1994). However, most research into the cause of injury and subsequent disablement experience is based on single dimensional, medical models that seek to explain injury in terms of individual workers' traits or simplistically theorized and measured work-environment factors. Increasingly, evidence indicates both that a number of workplace culture characteristics (such as senior management "buy-in" to health and safety and the extent to which workers participate in decision making) contribute to both injury incidence (Habeck et al, 1991; Shannon et al, 2000) and subsequent disability experience, including return to work (RTW) (Amick et al., 2000). This evidence points to the need for more comprehensive programs that take into consideration complex work environment factors to address both primary prevention (injury incidence) and secondary prevention (the disablement process) outcomes.

Development of more comprehensive models, particularly when these combine both primary and secondary prevention measures, may also break down the current isolation in research, policy making, and at worksites between injury prevention and RTW. Primary prevention of injury, identifying risk factors and addressing these prior to the occurrence of injury, is acknowledged to be the most humane approach to preventing disability from workplace injuries. Nevertheless, considerable controversy exists as to which primary interventions work best to prevent injuries (ergonomic alterations focussing on reducing biomechanical loads, addressing psychosocial factors, improving training, etc.), and, at least until recently, there has been a dearth of evidence on the economic benefit of injury prevention in relation to its costs (Norman and Wells, 2000). In this context, some employers have turned their focus to intervening only after an injury occurs, a secondary prevention approach, in attempting to reduce the burden of work-related disability and its costs.

In many workplaces and policy arenas, the professionals and stakeholders working to address "safety issues" are different from those who attend to the needs of injured workers. Ergonomists, safety personnel and physiotherapists generally assess the workplace and/or teach proper body mechanics and lifting techniques;

while claims management activity is generally conducted by physicians, nurses, resources personnel, and, professional claims managers. Often these two groups constitute two solitudes, and do not regularly interact. Furthermore, worker representatives (and/or unions, where they exist) are generally active in promoting the prevention of injury, but are often reluctant to support RTW programs. They fear that such programs will be used to harass workers to RTW before they are able, thereby risking further injury or re-injury.

While the themes of this chapter are applicable to all industries, the data and examples provided in this chapter come mainly from the healthcare sector. Injury rates are very high in this sector, as are the costs. According to the Workers' Compensation Board (WCB) in British Columbia (BC), between 1997 and 2000 there were more than 28,000 time-loss injuries to workers, with a rate of approximately 7.4 injuries for every 1,000 workers in the BC healthcare sector, compared to a rate of 4.8 in the province overall. Those injuries resulted in almost 1.5 million days lost (5,800 full time equivalents (FTEs) in total or 1,450 on average per year). The healthcare sector accounts for 10% of all time-loss claims in the province, more than any other industry. Direct claims costs for all injuries in the three-year period from 1997 to 1999 amounted to $180 million (Healthcare Industry Focus Report on Occupational Injury and Disease, 2000). In Canada overall, nursing personnel miss more time from work due to illness or injury than workers in any other sector (Labour Force Surveys, 1999; Akyeampong, 1999).

The BC WCB had attempted educational and other prevention campaigns targeted specifically at the healthcare sector, and recently articulated ergonomic regulatory requirements (WCB of B.C., 1999). Nonetheless, the overall decreases in injury rates and time loss that has occurred in other jurisdictions and in other sectors (Ostry, 2000) had not yet materialized. Thus, whatever is responsible for bringing down injuries elsewhere has not been sufficiently defined and/or operational in BC's healthcare sector. Similarly, efforts targeted at decreasing time loss by a more coordinated approach to the provision for injured workers – what has been known as "continuum of care" – have not been effective, nor have efforts to provide clinicians with guidelines as to what is reasonable time off work according to the nature of an injury – known as "length of disability guidelines" (Reed, 1997). Thus the considerable investments in claims management by the WCB have also not had the results envisioned, especially with respect to the costs of time loss claims in the BC healthcare sector. In fact, for the year 2000, the BC WCB was forced to write off a deficit of $9 million in the account of the healthcare sector, as compared with a surplus of $70 million three years earlier. To maintain balance in the account, WCB premiums for the healthcare sector were raised by 40% for 2001 ($25 million) and a further increase of 40% is expected for next year. Current rates are about 2% of assessable wages, and are likely to increase to between 2.7% (acute care) and 4.25% (long-term care) over time unless injury rates and injury costs decrease considerably (Perrin, 2001).

In a study funded through the Health Evidence Application and Linkage Network (HEALNet), "Work Ready" (Friesen *et al.*, 2001; Guzman *et al.*, 2003) researchers found that the major problem underlying failure to RTW was lack of communication and systemic organizational-level problems. As well, in a subsequent study, Guzman and colleagues found that doctors believed that failure to provide adequate accommodation in the workplace was among the most important factors delaying timely RTW (Guzman *et al.*, 2001). Nonetheless, in a survey of all healthcare facilities in BC (Ostry *et al.*, 2000) it was found that 78% of workplaces had policies and/or programs in place to reduce injuries, but less than 50% had any kind of RTW program.

This chapter will discuss the evidence suggesting that both primary and secondary prevention outcome can be positively influenced by programs focusing on either primary or secondary prevention, especially if workplace cultural issues, worker participation and senior management "buy in" are explicit. Linking primary and secondary prevention activities in this regard provides the most progressive means for decreasing injury-induced disability in the workplace; this linkage may not only reduce costs of operating two separate programs, but it should simultaneously enhance their mutual effectiveness. Indeed failure to link the two approaches may cause the loss of critical support needed from the proponents of each of the perspectives.

The chapter begins by outlining a concrete example from the BC healthcare sector, of a primary prevention (ergonomic) intervention that had a secondary prevention spin off as measured by reduced claim duration and decreased cost per claim. The next section of the chapter will look at the converse – how a secondary prevention effort had a primary prevention spin off: decreasing the frequency of injuries. In both cases, the success of the intervention lay in the extent to which worker participation was promoted and senior management support was available. Data from both examples will be situated within a review that suggests that work organization and a workplace culture of worker participation and senior management support are paramount to both.

4.2   PRIMARY PREVENTION EFFORTS LEAD TO SECONDARY PREVENTION SPIN-OFFS

Although relatively recent, there are an increasing number of examples in the published literature of ergonomic interventions that have been associated with decreased costs related to injuries through reductions in total absenteeism and total time loss (Lanoie and Tavenas, 1996). However, it is difficult to know whether such results stem solely from primary prevention activities, or whether there had also been a decrease in disability post-injury due to secondary prevention measures (as can be better judged through decreased duration or average cost per claim in association with targeted workplace accommodation measures). In a cost-benefit analysis of an ergonomic intervention in Comox, BC, we studied this specifically.

A "Resident Lifting System Project" was initiated in the Extended Care Unit of St. Joseph's Hospital in Comox through a grant of $344,323 from the WCB of BC. The objective of this project, which commenced in April 1998, was to reduce musculoskeletal injuries (MSI) to staff, specifically the injuries caused by lifting and transferring the residents in this unit. The major component of the project involved installing 65 mechanical ceiling lift devices, fitted within pre-existing structures, in all patient bed and bathing rooms. The ceiling lifts replaced a traditional floor lift system. A no-manual-lifting policy was implemented, with staff training provided in using the new devices. Staff were actively involved in all phases of this project, with extensive worker participation in the selection of the lifts themselves, as well as in all phases of the implementation, and even in the dissemination of the results.

An evaluation of this Resident Lifting System Project was conducted by the Occupational Health & Safety Agency for Healthcare (OHSAH),[1] a joint union-management governed agency, committed to decreasing injuries and time loss in healthcare workers. Injury data were gathered through analysis of injury reports and WCB claims, and costs and benefits attributable to the Resident Lifting System Project were identified and measured for a one-year period preceding and following the intervention. Direct benefits were calculated as the change in MSI-related compensation claims between the two periods. Assumptions regarding time preference (discounting the value of future costs and benefits against present values) and trends in injury rates were explicitly identified, and sensitivity analyses were carried out to assess the influence of changes in assumptions. Payback periods, benefit cost ratios and internal rates of return were calculated from the perspectives of both the insurer and employer.

While the incidence of "lift/transfer" claims decreased by 58% (from 24 to 10, p=0.01) (Ronald *et al.*, 2002) (the primary prevention payoff), the costs per 100,000 hours worked were reduced by 69% (from $65,997 to $20,731). The net reduction in the cost of claims filed in the study period following the intervention was $59,282 for lift/transfer MSI-related claims exclusively and $89,378 for all MSI-related claims. Taking this latter figure, a payback period of 3.85 years was estimated for the investment. Over the estimated 12-year lifespan of the equipment, the present value of the accumulated claims cost reductions exceeded the investment cost by a factor of 2.5 to 1, representing an internal rate of return of 8.1%. From the perspective of the facility itself, the present value of all direct and estimated indirect benefits exceeded that of all costs associated with the intervention by a factor of 6.1 to 1, representing an internal rate of return of 17.9% (Spiegel *et al.*, 2001). The savings that accrued came from both reduced MSI incidence and reduced duration of claims. As there had been no change in claims management procedure or RTW program, it can be deduced that the reduced time loss per claim thus occurred mainly because of the primary intervention.

Now, successful implementation of a primary prevention intervention itself follows from two organizational considerations: the ultimate decision to proceed and the culture accompanying its introduction. In the fall of 1999 and spring of 2000, OHSAH conducted a needs assessment of over 450 BC healthcare facilities by surveying labour and management representatives on the Joint Health and Safety Committee (JHSC) at each of these facilities. A validated set of questions assessing "senior management buy-in to health and safety" was part of this survey (Ostry *et al.*, 2000). The average score (on a scale from 1 to 100) for senior management buy-in to health and safety for BC facilities was 73.4 compared to a score at this facility of 95.0, validating worker observations accompanying the Resident Lifting System Project that "management cared" (Yassi *et al.*, 1995a,b). While management's willingness to pursue the ergonomic intervention itself was undoubtedly associated with the high score awarded, the manner in which the project was introduced appears itself to have contributed to successful implementation. Specifically, prior to full implementation of the project in the Extended Care Unit at the Comox facility, a ceiling lift was first piloted in a single room, and staff were directly involved in evaluating its effectiveness and the feasibility of broader deployment. Thus the decreased time-loss per claim may reflect less severe injuries, allowing a more prompt RTW. However, it was felt in part to be due to a more welcoming physical workplace, one that required less lifting, a known major factor preventing RTW (Cooper *et al.*, 1998). Moreover, the involvement of the workforce in the implementation of this primary prevention intervention itself changed the culture of the workplace, making it more socially welcoming, and likely also played a major role in decreasing time loss and costs.

4.3 SECONDARY PREVENTION EFFORTS LEAD TO PRIMARY PREVENTION SPIN-OFFS

Almost a decade earlier, an intervention study was conducted over a two-year period beginning on October 1, 1990, at the Health Sciences Centre (HSC), a large acute and tertiary care teaching hospital in Winnipeg, Manitoba. After extensive discussion with senior management and labor leaders, it was decided that the intervention should be targeted to those units that needed it most, and that the focus should be on workplace accommodation as soon as possible after an injury. Hospital wards were therefore designated either as high risk or lower risk for back injury, based on ergonomic assessments and review of previous injury data. As agreed, study wards were the highest risk wards, and these were targeted for a program that stressed early intervention post injury, with provision for modified work. According to the protocol developed through extensive consultation, all injured nurses were contacted as soon as possible after the reported injury (Yassi *et al.*, 1995a), with every effort being made to interview nurses within two working days following the injury. The injured nurse was asked a series of questions as to the mechanism of the injury (Yassi *et al.*, 1995a), as well as what could have been done to prevent the injury. With this in mind, efforts were then made to aggressively identify options for modifying the injured worker's tasks.

The results were markedly positive in all respects. There was a 29% decrease in time lost per 100,000 paid hours compared to prior to the intervention in the target groups, while an increase occurred on the wards not targeted for the intervention; there was significantly greater reduced pain and disability six months following the injuries on the target wards in comparison to the non-targeted wards; and no one from the targeted wards who agreed to enter the program was still off work six months following the injury, compared to six individuals in the non-targeted wards. Most astonishingly, however, was that there was a 33% reduction in the frequency of injuries (and 44% reduction in time-loss injuries) on the targeted wards, again while there was a marked increase in the injury rate on the non-targeted wards (Yassi *et al.*, 1995a).

It appears that the primary prevention spin-off benefit observed in this study was due to both generalized and specific processes set in motion by implementation of the secondary prevention intervention (Yassi *et al.*, 2000). The early intervention program began by extensive discussions about this issue with workers and management, as well as a campaign raising awareness about injuries and their mechanism among all staff in targeted wards in order to promote early reporting post-injury so that immediate follow-up can occur. This increased awareness may have contributed to the primary prevention results by increasing safety awareness in general. Additionally, in the immediate injury follow-up targeted to occur within 24 to 48 hours post-injury, the injured worker was interviewed as to the cause of the injury and what might have prevented it, as well as what was needed to accommodate the worker back in the workplace. In the course of accommodation, changes were made to the workplace or, more often than not, to work task organization, again with the main objective of returning the injured worker to his or her regular work. Clearly, however, other workers benefit from these same changes, whether these are work alterations, greater awareness of proper lifting technique, or increased availability of equipment. While the case described from BC was one in which intervention aiming at primary prevention led to decreased time-loss per claim without a specifically designed secondary prevention intervention, in this case, an intervention aimed at secondary prevention had a primary prevention affect as well.

To better understand the determinants of time loss, three statistical modeling techniques were then used to specifically examine the occurrence of time loss, duration of time loss, and duration of time loss once an injury incurring lost time had been documented (Tate *et al.*, 1999). First, the logistic regression model was used to relate characteristics at the time of injury to a binary indicator of any time-loss after the injury. Second, the duration of time loss was modeled using the Tobit regression model. This model accounts for censoring of time loss at zero, appropriate in situations in which there is an imposed lower limit to the dependent variable. Third, ordinary least-squares linear regression was used to examine the factors related to duration of time loss in nurses who incurred at least one day of time loss. The findings showed that the perception of disability at the time an injury occurred was critical to prevention of time loss. This may be indicative: with a reassuring message that the injury need not be "disabling" – for example, with the knowledge that the worker can be accommodated at work – the need for time loss is prevented. However, the time loss findings show that once time loss has occurred, attention to providing modified work and attention to pain reduction are highly warranted. The fact that the offer of an early intervention program that includes modifications to the workplace or work assignments was a key factor in reducing the duration of time loss added to the evidence that secondary prevention workplace intervention programs can be effective.

4.4 WORK ORGANIZATIONAL FACTORS LED TO BOTH PRIMARY AND SECONDARY PREVENTION SPIN-OFFS

It is now well documented that the most successful intervention programs aiming to reduce workplace health and safety risks are those that target administrative and management procedures, taking a more comprehensive approach not only to primary prevention of injuries but also to secondary prevention (Frank *et al.*,

1996). Habeck *et al.* (1991) have developed and tested a model that links the organizational characteristics of a firm with injury incidence, disability incidence and disability costs. In their model, broad non-occupational health and safety (OH&S) characteristics of the organization such as its size, industry type, level of unionization as well as demographic characteristics of the workforce (such as level of seniority, turnover rate, and average age) have been shown to influence injury rates at the firm level. While these are not particularly amenable for intervention, organizational characteristics of firms in this model also included "managerial style and firm culture". These less tangible facets of an organization were shown to shape the work climate, and were a measure of the organization's relationship with its employees. According to Amick, these organizational characteristics "are subject to change over time" and are therefore presumably modifiable and therefore potentially a focus for interventions to improve injury outcomes in a firm or institution (Amick *et al.,* 2000).

The model also includes firm-level policies and practices directly related to OH&S – theorized by researchers as being nested within the more distal organizational characteristics just described. Firm-level policies and practices are divided into primary prevention activities of the firm, such as policies, practices or interventions directed to changing individual safety behaviors on reducing exposures, and secondary prevention policies, practices or interventions that occur after injury. These latter policies might include physical rehabilitation, disability case management, or early RTW programs.

In a comprehensive review of the literature relating firm-level OH&S policies, practices and interventions with injury outcomes, the conclusion was drawn that, among several organizational-level OH&S factors, "perhaps the most important among them is the commitment of top management to safety" (Shannon *et al.,* 2000). Regardless of the type of study, the finding that such commitment is important is supported repeatedly in study after study. This important firm-level characteristic is thus an element of "management style and firm culture", located more specifically within the OH&S organizational characteristics of the firm, but also presumably embedded within the non-OH&S organizational culture.

Amick *et al.* (2000) have taken Habeck's and colleagues (1991) model several steps forward. First they used worker respondents to characterize organizational level and OH&S policies at their workplaces. Most researchers investigating organizational characters in relation to injury have relied on reports of management – usually human resources managers. Second, they conducted factor analyses in order to better characterize Habeck *et al.*'s scale, reducing the number of items in order to make the instrument more manageable. And, third, they used this refined model to predict "RTW" for workers off work with carpal tunnel syndrome. In this investigation they demonstrated that "people-oriented culture", a "safety climate", good ergonomics procedures and the presence of "good" disability management programs independently predicted RTW six months after carpal tunnel injury. Thus, the work Habeck *et al.* and Amick show that non-OH&S characteristics of work organization impact both injury incidence and outcomes subsequent to injury.

Recent work with this model points to the importance of organizational characteristics and OH&S interventions in predicting injury incidence, disability incidence and RTW. There is, however, a need to further explore the link between the non-OH&S organizational factors, the organizational factors that include

OH&S (such as senior management commitment of OH&S), and both primary and secondary OH&S policies, practices and interventions, as well as to improve injury/RTW outcomes in order to identify the "best" places to intervene. Until these links are explored further it may be useful to review the relatively small literature on interventions designed to alter organizational structure in order to improve firm-level health outcomes.

A review was done of interventions designed to produce organizational change aimed at improving worker health outcomes through identified studies (Polanyi *et al.*, 1999). Most of these were in American manufacturing firms with 100 or more employees. Nearly all of the interventions involved organizational changes, such as development of semi-autonomous work teams; enhanced staff-management interaction, in conjunction with a career advancement scheme; a participatory action research plan that brings together academics, unions, and management decision-makers to improve the organizational climate; and enhanced worker participation facilitated through a participatory research action plan. Polanyi's review showed mixed results, although "a frequent refrain was the need for early, wide, and significant employee and management participation in interventions". Furthermore, "participation is, by definition, key to action research projects, and strong participation also often leads to sustainability".

It has been shown that work organization factors, including high workloads and limited control over how work is performed, influence the incidence of low back problems among nurses (Niedhammer *et al.*, 1994; Josephson *et al.*, 1997; Koehoorn *et al.*, 1999). Inadequate staffing, with its impact on insufficient staff for proper lifting techniques, has been specifically identified (Koehoorn *et al.*, 1999; Larese and Fiorito, 1994). The influence of these factors, on rates of injuries and duration of time loss, is less well understood, but it has been shown that the extent to which a hospital organisation encourages staff to use creative strategies to cope with difficult workloads influences both injury and time loss outcome (Estryn-Behar, 1998). Other programs in healthcare that have indeed been successful in reducing musculoskeletal disorders and subsequent time loss, similar to the results shown for the above cases, have been those that adopted a multi-faceted approach (Garg and Owen, 1992; Nordstrom-Njorverud and Mortiz, 1998; Wood, 1987). In spite of this, it should be noted that healthcare facilities might be implementing fewer new management ideas than most industries, according to a Swedish survey (Le Grand and Szulkin, 1993).

4.5 CONCLUSION

Workplace studies suggest that primary prevention (reduction of injuries) can emerge as an outcome of a well-designed "RTW" program; that RTW can, in turn, be promoted by ergonomic and work organizational changes; and that workplace culture featuring active worker participation and senior management support encourages both injury prevention and reduced time loss. In recognition of this evidence, organizations should look at designing and implementing comprehensive programs that address both these issues. The recent research on organizational culture as a major determinant of injury rates in general, and as an important predictor of RTW, indicates that the likelihood of successful implementation of such comprehensive programs will be higher within more functional organizational cultures.

OHSAH, which evaluated the Resident Lifting System, is an unusual organisation. It is bi-partite, with a joint labor-management governing structure that fosters worker and employer "buy-in" and involvement, from the OHSAH Board members through to joint health and safety committees at the individual facilities. This provincial-level organization embodies and actively models organizational level attributes (usually discussed in the literature in relation to individual facilities only), such as worker-management participation and senior management buy-in to health and safety. OHSAH, which actively models dimensions of organizational culture that have been linked to improved primary and secondary outcomes at the facility level, is likely very useful in both promoting the adoption of such values and in having a local impact at the project implementation level.

OHSAH is sponsoring integrated prevention and RTW programs in the healthcare sector in the province of BC. To obtain funding for these projects a workplace must agree to a set of explicit principles such as a focus on workplace accommodation, provision of meaningful alternate or modified work assignments, and insuring a strong component of prevention and ergonomic intervention to facilitate RTW. This contrasts with a "medical model" approach that exclusively concentrates on rehabilitation of the injured worker. In addition, the program requires a commitment to broader organizational level values and actions such as senior management commitment, both financially and through active participation, as well as extensive worker participation throughout the process, from the design of the intervention program to its evaluation.

Experiences in Winnipeg and BC's healthcare sectors, documenting explicit links between primary and secondary prevention, argues in favor of breaking silos down and adopting a comprehensive approach with senior management commitment and meaningful worker input as key to a healthy workplace.

4.6 KEY MESSAGES

- Primary prevention measures, if implemented properly, should result not only in decreased injuries, but also decreased duration of time loss post injury (i.e. have a secondary prevention spin off).

- Secondary prevention measures – i.e. early intervention post-injury, if focused on workplace modifications to accommodate the injured worker, should result not only in decreased time loss post-injury, but also fewer new injuries (i.e. have a primary prevention spin off).

- Both rates of injuries and duration of time loss post-injury are strongly influenced by workplace culture and organizational factors. Thus efforts to improve workplace culture and work organization should result in both decreased injuries and decreased time loss.

- Senior management commitment and meaningful worker participation are crucial to preventing injuries and time loss.

- Designing and implementing a comprehensive program that breaks down the silos between those who are charged with preventing injuries and those who assist the injured worker to RTW makes good sense.

4.7 NOTES

[1]OHSAH was established in 1999 through collective bargaining in the BC healthcare sectors with a mission to define best practices, pilot-test, evaluate then generalize results (Lanoie and Tavenas, 1996; Ronald *et al.*, 2001).

4.8 REFERENCES

Akyeampong, E.B., 1999, *Missing work in 1998: Industry differences*, (Ottawa: Statistics Canada).

Amick, B.C.3rd, Lerner, D., Rogers, W.H., Rooney, T. and Katz, J.N., 2000, A review of health-related work outcome measures and their uses, and recommended measures. *Spine*, **25**, pp. 3152-3160.

Cooper, J.E., Tate, R. and Yassi, A., 1998, Components of initial and residual disability following back injury in nurses. *Spine,* **23**, pp. 2118-2122.

Estryn-Behar, M., 1998, Ergonomics and health care. In: ILO, ed. *Encyclopaedia of Occupational Health and Safety.* Geneva: ILO, 1998, pp. 97.26-97.33.

Frank, J.W., Kerr, M.S., Brooker, A.S., DeMaio, S.E., Maetzel, A., Shannon, H.S., Sullivan, T.J., Norman, R.W. and Wells, R.P., 1996, Disability resulting from occupational low back pain. Part II: what do we know about secondary prevention? A review of the scientific evidence on prevention after disability begins. *Spine*, **21**, pp. 2918-2928.

Friesen, M., Yassi, A. and Cooper, J., 2001, The importance of human interactions and organizational structures. *Work Journal*, **17**, pp. 11-22.

Garg, A. and Owen, B., 1992, Reducing back stress to nursing personnel: an ergonomic intervention in a nursing home. Ergonomics, 34, pp. 1353-1375.

Guzman J., Frank J., Stock, S., Yassi, A. and Loisel, P., 2003, Stakeholder Views of Return to Work after Occupational Injury. In: Sullivan, T. and Frank, J. editors. *Preventing and Managing Disabling Injury at Work*, (London: Taylor & Francis). Chapter 5, in this book.

Guzman, J., Cooper, J.E., Khokhar, J. and Yassi, A., 2001, Perspectives of Primary Care Physicians on RTW after a soft tissue injury. Submitted to *Canadian Family Physician*, 2001.

Habeck, R.V., Leahy, M.J., Hunt, H.A., Chan, F. and Welch, E.M., 1991, Employer factors related to workers' compensation claims and disability management. *Rehabilitation Counselling Bulletin*, **34**, pp. 210-226.

Healthcare Industry Focus Report on Occupational Injury and Disease, 2000, BC WCB.

Josephson, M., Lagerstrom, M., Hagberg M. and Hjelm, E.W., 1997, Musculoskeletal symptoms and job strain among nursing personnel: a study over a three year period. *Occupational and Environmental Medicine*, **54**, pp. 681-685.

Koehoorn, M., Kennedy, S.M., Demers, P.A., *et al.*, 1999, Work organisation factors and musculoskeletal outcomes among a cohort of health care workers. *Health Care & Epidemiology.* (Vancouver: University of British Columbia)

Labour Force Surveys, 1999, Statistics Canada.

Lanoie, P. and Tavenas, S., 1996, Costs and benefits of preventing workplace accidents: the case of participatory ergonomics. *Safety Science*, **24**, pp. 181-196.

Larese, F. and Fiorito, A., 1994, Musculoskeletal disorders in hospital nurses: a comparison between two hospitals. *Ergonomics*, **37**, pp. 1205-1211.

Le Grand, C. and Szulkin, R.M.T., 1993, Swedish work places – organisation, personnel development and management. In: How to deal with stress in organisations? – a health perspective on theory and practice, edited by Theorell T., (City: press), pp. 616-624.

Niedhammer, I., Lert, F. and Marne, M., 1994, Back pain and associated factors in French nurses. *International Archives of Occupational and Environmental Health,* **66**, pp. 349-357.

Nordstrom-Njorverud, G. and Mortiz, U., 1998, Interdisciplinary rehabilitation of hospital employees with musculoskeletal disorders. *Scandinavian Journal of Rehabilitation Medicine,* **20**, pp. 31-37.

Norman R. and Wells R., 2000, Ergonomic interventions for reducing musculoskeletal disorders. In: Sullivan, T. (ed.) *Injury and the New World of Work*. (University of British Columbia Press.)

Ostry, A., 2000, From Chainsaws to Keyboards: Historical Trends in the Epidemiology of Injury and Industrial Disease in British Columbia, (1950-1996). In Sullivan, T. (ed). *Injury and the New World of Work*. (Vancouver: UBC Press).

Ostry, A., Yassi, A. and Tate, R., November 2000, A Needs Assessment for Occupational Health and Safety Programs in British Columbia's Healthcare Sector: A Joint Labour-Management Approach. Presented at *128th American Public Health Association Annual Meeting*, Boston, Massachusetts.

Perrin, D., April 2001, *Financing Options for Patient Lift and Transfer Equipment in the BC Health Care Sector.* Report prepared for the Occupational Health and Safety Agency for Healthcare in British Columbia.

Polanyi, M., Eakin, J., Frank, J.W., Shannon, H. and Sullivan T., 1999, Creating Healthier Work Environments: A Critical Review of the Health Impact of Workplace Change. In Vol.3, *Canada Health Action: Building on the Legacy.* Papers commissioned for the National Forum on Health. Ottawa, Ontario.

Reed P., 1997, *Medical Disability Advisor: Workplace Guidelines 3rd Ed.* (Reed Group Ltd).

Ronald, L.A., Yassi, A., Tate, R.B., Siegel, H., Tait, D. and Mozel, M.R., 2002, Effectiveness of installing overhead ceiling lifts on reducing musculoskeletal injuries in an extended care hospital unit. *American Association Occupational Health Nurses Journal*, **50** (3): pp. 120-127.

Shannon, H.S., Woodward, C.A., Cunningham, C.E., McIntosh, J., Lendrum, B., Brown, J. and Rosenbloom, D., 2000, Changes in general health and musculoskeletal outcomes in the workforce of a hospital undergoing rapid change: a longitudinal study. *Journal of Occupational Health Psychology,* **6**, pp. 3-14.

Spiegel, J., Yassi, A., Tate, R.B., Tait, D. and Ronald, L.A., 2001, Implementing a resident lifting system in an extended care hospital. *American Association of Occupational Health Nurses Journal,* **50** (3): 128-134.

Tate, R.B., Yassi, A. and Cooper, J., 1999, Predictors of time loss after back injury in nurses. *Spine,* **24**, pp. 1930-1935.

Verbrugge, L.M. and Jette, A.M., 1994, The disablement process. *Social Science Medicine,* **38**, pp. 1-14.

WCB of B.C. (1999). Ergonomic Draft Operating Instructions. Richmond, B.C., *Worker's Compensation Board of British Columbia,* B4-1 to B4-13.

Wood, D., 1987, Design and evaluation of a back injury prevention program within a geriatric hospital. *Spine,* **12**, pp. 77-82.

Yassi, A., Tate, R., Cooper, J.E., Snow, C., Vallentyne, S. and Khokhar, J., 1995a, Early intervention for back-injured nurses at a large Canadian tertiary care hospital: an evaluation of the effectiveness and cost benefit of a 2-year pilot project. *Occupational Medicine,* **45**, pp. 209-214.

Yassi, A., Khokhar, J., Tate, R., Cooper, J., Snow, C. and Vallentyne S., 1995b, The epidemiology of back injuries in nurses at a large Canadian tertiary care hospital: Implications for prevention. *Occupational Medicine,* **45**, pp. 215-220.

Yassi, A., Cooper, J. and Tate, R.B., 2000, Letter to Editor; 2000. Early intervention for back-injured nurses at a large hospital. *Spine,* **25**, pp. 2549-2553.

CHAPTER 5

Stakeholder Views of Return to Work after Occupational Injury

Jaime Guzman, John Frank, Susan Stock, Annalee Yassi and
Patrick Loisel for the Work-Ready Group[1]

5.1 BACKGROUND

In 1995 alone, more than 400,000 Canadians lost time from work due to work injuries and the Workers' Compensation Boards in Canada paid close to five billion dollars in benefits. With the addition of the indirect costs, the annual total cost of occupational injuries to the Canadian economy is estimated to be close to $9.9 billion (Liberty International Canada, 1995). Most occupational injuries are soft tissue injuries such as occupational back pain, sprains, strains and tendonitis. In fact, a developed country could spend more than 1% of its gross national product in dealing with the direct and indirect costs of back pain alone (van Tulder et al., 1995).

   Most costs associated with work injuries are a direct consequence of time lost from work. Many persons and groups have a stake on the timely return to work (RTW) of injured workers. The workers themselves, and their representatives, would like to minimize loss of income and chances of re-injury (Baril et al., 1994). Employers face disruptions in production and increases in their worker compensation insurance premiums. The insurer is directly liable for compensation benefits (Hager, 1993). Providers of healthcare and rehabilitation services are increasingly held accountable for the RTW outcomes of workers under their care.

   Many of these stakeholders also have a direct or indirect influence on the timing of RTW. It is an occupational health premise that all stakeholders ought to be involved for RTW programs to work (Shrey, 1996; Frank et al., 1998), yet few scientific examinations of stakeholder perspectives have been published (Brines et al., 1999; Pergola et al., 1999). Understanding of stakeholder perspectives is essential for motivating and sustaining their involvement in RTW (RTW) and in preventing disability (Frank et al., 1996; Frank et al., 1998). To better understand stakeholders' perspectives, the authors collaborated in a multidisciplinary research project named Work-Ready (Work-Ready Research Group, 1999).

   The project involved researchers in occupational health and rehabilitation in three Canadian provinces brought together under the auspices of the Canadian Networks of Centres of Excellence initiative (NCE). Work-Ready researchers are part of HEALNet, an NCE-funded research network devoted to enhancing the health of Canadians through improved use of the most relevant research evidence in health decision-making. HEALNet stands for Health Evidence Application and Linkage Network. The researchers and institutions involved in the Work-Ready project are listed at the end of the chapter.

The Work-Ready project had two objectives: first, to describe stakeholder views of factors influencing RTW after occupational soft tissue injuries and second; to develop tools to facilitate constructive stakeholder dialogue around available scientific evidence and experience on RTW issues. Work-ready researchers chose to use qualitative research methods to describe stakeholder perspectives. The tools for dialogue were developed around "workshops", face to face encounters of diverse stakeholders to discuss case scenarios of workers with occupational injuries. This chapter summarizes what Work-Ready researchers learnt about the perspectives of diverse stakeholders on RTW after soft tissue injuries. The key points on the development of the Work-Ready "tool-kit" are listed at the end of the chapter.

5.2 FIELDWORK METHODS

Fieldwork consisted of three parallel studies in Manitoba, Ontario and Quebec. They had the same general objectives and approach, but study methods were adapted to the specific contexts in each of the three provinces. The goal was to gather occupational stakeholders' perspectives on barriers and facilitators for RTW after soft tissue injury, and on effective strategies and/or tools for RTW. Individual interviews, focus groups and some written questionnaires were used. Two main questions were asked:

- What have you found to be the challenges/barriers in facilitating worker recovery and RTW after (soft tissue) injury?

- What solutions have you tried/observed that work?

In Manitoba, researchers interviewed a total of 55 individuals, either in one-on-one encounters or in small groups, from ten workplaces, four provincial agencies and two professional groups. Interviewees included injured workers, union representatives, managers, RTW coordinators, healthcare providers, vocational rehabilitation specialists and insurer representatives. The spectrum of workplaces represented health services, public services, construction, retail, manufacturing and hydro-electric services (Friesen *et al.*, 1999).

In Ontario, most data were derived from 17 semi-structured interviews in 11 employment settings in two mid-sized cities. Workplaces included manufacturing firms, hospitals and one municipal government. Interviewees consisted predominantly of occupational health nurses, and included union representatives, health and safety coordinators, one physician, a physiotherapist and a manager. Additional information was gathered from a workshop with local physicians and interviews with a variety of stakeholders regarding the cases of individual workers with soft tissue injuries, which evolved into chronicity (Clarke *et al.*, 1999). In Quebec, Work-Ready researchers interviewed 36 individuals in ten electric and electronics companies located on the island of Montreal. Interviewees included human resources and health and safety managers, medical services personnel, supervisors, union representatives, injured workers and two staff of the CSST (Quebec's Workers' Compensation Board) (Baril *et al.*, 2000).

Overall, employment settings ranged from the very small (i.e. less than 50 employees) to the very large (more than 5000 employees) and encompassed a wide variety of manufacturing and services firms, public and private. The collaboration between the groups in the three provinces allowed us to compare and contrast the ideas expressed by interviewees in these varied contexts (both cultural and legislative), and facilitated a more comprehensive understanding of the perspectives of various occupational health stakeholders about RTW. This "triangulation" of actors and settings further enhances the trustworthiness of the information obtained (Gliner, 1994).

When analyzing recorded interviews, comments were first divided into two large categories: things which people told us were helpful in returning the injured worker to the job, and those which acted as barriers in the process. As expected, issues were often raised which could be interpreted either way – for example, the presence of a workplace or worker characteristic was seen as a facilitator, while its absence was seen as a barrier to successful RTW. With the help of qualitative research software (ATLAS.ti and Q.S.R.NUD*IST), comments were further categorized by common themes and relationships between themes were analyzed and discussed. Throughout this chapter, direct quotes from interviewees are included to emphasize important points. As appropriate, we have identified the speaker in terms of being a worker, manager, healthcare professional and so on.

5.3 FIELDWORK FINDINGS

As we conducted the analyses of interview transcripts it became evident that, while a few barriers or facilitators for RTW could be clearly assigned to the attitudes and/or actions of one particular stakeholder group or the injured workers themselves, most of them in fact arose during interactions between stakeholders, either within the workplace or outside the workplace. In some instances the barriers or facilitators for the RTW process seemed to arise more or less directly from the regulatory and social-economic environment in the province. The main barriers and facilitators identified by our interviewees are described in the following three broad categories: interactions within the workplace, interactions with stakeholders external to the workplace, and the social, economic and regulatory environment (Figure 5.1).

5.3.1 Interactions within the Workplace

Interactions of injured workers with co-workers, supervisors, occupational health providers and RTW managers within the workplace are perhaps the most important factors determining the timing and success of RTW after a soft tissue injury. The frequency and tone of these interactions are determined by the workplace culture of the firm, understood as the shared key values of people at the workplace which guide and shape behavior (Smircich, 1983). Workplace culture includes aspects of organization and production, which provide the group with mechanisms to interact with others and the environment (Krefting and Krefting, 1991).

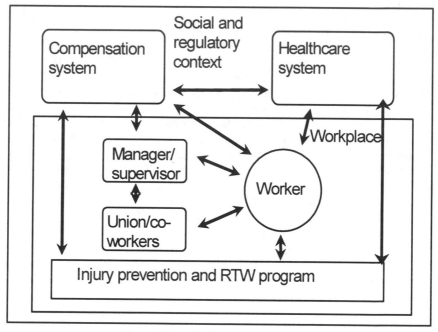

Figure 5.1 Levels of interaction in RTW.

The workplace culture in turn determines the goals and objectives of RTW programs. In the workplaces studied, program goals could include reducing Workers' Compensation Board (WCB) costs, increasing productivity of workers with soft tissue injuries, avoiding recurrences and exacerbations of the injury, or preventing injuries altogether. Programs may officially subscribe to several of these goals, but our informants often perceived that one of them was clearly dominating. The perceived motives of the employer for instituting an RTW program were fundamental in building the trust of workers and other stakeholders. This trust was widely seen as essential to the eventual success of the program.
The following are elements of the workplace culture, which seemed to have a major impact on the success or failure of the RTW process.
The supportive or non-supportive supervisory and co-worker attitudes towards the injured worker. RTW was problematic in settings where injured

workers were viewed with suspicion and considered "cherry pickers" looking for a less demanding job.

Labor relations – in unionized workplaces, whether the union was viewed as a facilitator or barrier – depended on whether or not the union constructively participated in RTW plans and programs, and whether injured workers felt supported or not supported by their union.

Researchers also identified the facilitating effect of a participatory management style, characterized by the ability to integrate the points of view of workers and front-line supervisors in planning and implementing RTW activities.

Some elements of the organizational structure of the firm such as hierarchies and accounting structure also have an impact on the outcomes of the RTW process. The presence of a rigid hierarchy was sometimes seen as counteractive to the flexibility necessary to accommodate injured workers. Whether the production department or unit was or was not responsible for the cost of the injury's lost-time compensation was occasionally viewed as influencing whether or not that department was open to having the injured worker come back on modified duties.

Some aspects of work organization also had an impact on the success of RTW programs. Production demands could pose a problem in the success of RTW programs. These demands could come from officially sanctioned production quotas or indirectly from company set rewards (financial or recognition). Alternatively, shop floor culture may encourage the maintenance of high production levels ("We're going to do 1200 units a shift – we always have!"). In many instances supervisors experienced considerable role conflict between their production responsibilities and their responsibilities to accommodate injured workers. Some stakeholders interviewed felt that workplace organization into teams or cells facilitated the RTW process.

Worker attitudes and behaviors were perceived as important facilitators or barriers, both by the worker and by other stakeholders. Lack of motivation or the presence of pre-injury work performance problems was viewed as a barrier to RTW by many managers and professionals responsible for RTW programs. Workers spoke in terms of it being helpful to have a positive attitude toward all of life and work. A worker's domestic situation and social relationships on the job were also considered important – perhaps more than other characteristics such as age, general health status, or severity of injury.

Workers described feelings of "dis-empowerment". These feelings pertained to both workplace interactions and to interactions outside the workplace. Workers perceived an imbalance of power in their interactions with the disability insurer, in that the insurer had the power to withhold financial benefits; and the employer, in that the employer had the power to deny the worker a job. Often the worker felt incapable of functioning within "the system"; the system variously referred to the insurance system, the healthcare system and/or the workplace. The system was perceived as too complex, too overwhelming, or having too much arbitrary power.

Some interviewees, including health and safety managers, workers, supervisors and union representatives, pointed out that worker attitudes and behaviors are often the result of the workplace culture and/or the organizational structure and work organization of the firm. For example, in workplaces where the employer provides comprehensive injury prevention services and early support along with RTW programs, administrative and medical follow-up after an injury are more likely to be perceived by workers as reflecting a genuine concern for their health and welfare, rather than as a single-minded preoccupation with cost-control. In such firms, workers may be more willing to collaborate with the RTW program.

Physical attributes of the workplace can also act as facilitators or barriers to RTW. Poor ergonomics are an obvious disadvantage; but even jobs designed to be "ergonomically correct" can lack the flexibility to allow returning workers to alter their position or pace themselves.

In regard to workplaces where stakeholders felt they had a successful RTW program in place, two positive attributes of the RTW coordinators were mentioned by some stakeholders. First, in some workplaces there was the perception that the success of the RTW program was the result of the coordinator's personality, vision and hard work. Second, RTW coordinators themselves often referred to a sense of "ownership" or "discovery" which was linked to the development of the program.

5.3.2 Interactions Outside the Workplace

One element stressed when addressing interactions with stakeholders outside the workplace was the need for open, honest communication. Although communication within the workplace was clearly identified as important, lack of communication with or between stakeholders outside the workplace was seen as the major external barrier to successfully arranging RTW.

Employer representatives and occupational health professionals frequently cited the difficulty they experienced in their attempts to communicate with injured workers' treating physicians. This might occur if work restrictions were not specified clearly, if clinical information necessary for claim processing was lacking, or if specific questions needed to be addressed before modified work was offered. Interviewees stated that some physicians responded well to requests and were most helpful in problem-solving around RTW, whereas others were consistently unavailable for consultation, or in extreme cases, hostile to any attempts at communication.

On the other hand, the potentially adversarial nature of the compensation claims adjudication process, and the serious negative repercussions for the worker that clinical judgments can have, were sources of concern for physicians in dealing with injured workers. Healthcare professionals are required to complete insurer forms to provide information on the injured worker's physical status, and restrictions which should be taken into consideration in planning RTW. This is a time-consuming process in a busy clinical practice. An additional problem was the difficulty expressed by physicians in answering the questions asked on these forms with any degree of certainty – questions whose answers could ultimately have financial repercussions for the worker. A good example is questions regarding the projected date of RTW. Community physicians themselves noted that they often "go by what the worker tells them". Ironically, recent research has indicated that the predictions made by the worker are indeed one of the strongest predictors of lost-time duration (Mondloch and Cole, 1999), and that expectations for recovery are associated with outcomes in different diseases (Mondloch *et al.*, 2001).

Another area where communication was felt to be crucial was the interaction of employer and employee with the disability insurer. The complexity of the compensation system was described as a source of potential problems by some interviewees. As with any large service organization, there were good and bad examples of interactions with insurance personnel – some adjudicators were felt to be arrogant or delayed the process by not returning phone calls, whereas some were effective in their job and helpful in the process of RTW.

Besides communication problems, a second important marker of unsuccessful interactions outside the workplace was lengthy delays. These were experienced by all stakeholders in areas such as relaying information or waiting for medical tests – it was stated that such delays often contributed to the development of prolonged secondary disability, often marked by chronic pain.

"Education" was often perceived to be a solution for successful RTW; interestingly, it was usually the perception of some stakeholders that certain other stakeholder groups would benefit from education on particular aspects of RTW!

5.3.3 The Social, Economic and Regulatory Context

Even though the three regions we studied were all part of the same country, there were significant differences across the three social, economic and regulatory contexts. These differences in context influenced the interactions among stakeholders, and acted as facilitators or barriers for RTW. Within a particular province some elements of context may even differ across firms. For example, the economic pressures and competition may be very different in the private electronics sector from the publicly funded healthcare sector.

The following are elements of the social, economic and regulatory context that seemed to be influencing the RTW process.

In response to economic pressures, many workplaces were outsourcing specialized operations and downsizing their workforce, thus making it more difficult to identify alternative or modified work.

Reimbursement patterns for primary care physicians in the provinces studied did not reward extended time spent counseling, or negotiating RTW for an injured worker.

During our study, new legislation had been introduced in Ontario but not yet passed. The WCB was slated to become less involved in the determination of work readiness and negotiating RTW. This responsibility was to be transferred to employer and worker, with health professionals' assistance.

In some jurisdictions the occupational disability insurer provided lucrative rebates to employers who had modified work placements. This often resulted in make-work placements, which were not meaningful to the worker.

Around the time of our study, the Canadian Medical Association and several provincial medical associations and medical regulatory bodies published position papers that were meant to guide the interaction of physicians with workers and employers. These documents argued for lessening the "adjudicatory" role of the physician in the determination of compensable disability and work-readiness. They also reinforced the physician's duty to preserve confidentiality, unless the information was specifically related to functional abilities and work-restrictions needed to plan RTW.

5.4 IMPLICATIONS

Despite differences in contexts across the three provinces (Manitoba, Ontario and Quebec), interviews with a variety of occupational health stakeholders support the importance of a number of facilitators and barriers for RTW after a soft tissue injury. The emergence of common themes despite a considerable diversity of interviewees and workplaces sampled, allows us to have confidence that the themes identified in this study are representative of the experiences of many Canadian workers and workplaces. The identified barriers and facilitators can be meaningfully arranged into three categories.

- Interactions within the workplace, including elements of the workplace culture, organizational structure and work organization.

- Interactions outside the workplace, particularly those involving the healthcare system and the disability insurance system.

- The characteristics of the social, economic and regulatory context which modulate all the interactions listed above.

Two qualities of the interactions seemed crucial in understanding the success or failure of the RTW of individual workers after a soft tissue injury. First, interactions characterized by honest and open communication were seen as facilitating RTW by most of the interviewed stakeholders. Second, interactions characterized by trust on the other stakeholder and commitment to the process, were seen as essential for successful RTW. The following sections discuss theoretical implications of our findings and list constructive suggestions for practitioners and researchers that we feel are firmly based on what we learned through our interviews with occupational health stakeholders. It should be pointed out, nonetheless, that at present there is a lack of rigorous evaluation studies supporting their efficacy as specific policy and program strategies.

5.4.1 Implications for Theory

One main premise of qualitative naturalistic research is that theory emerges from the data. To facilitate this process, Work-Ready researchers did not impose a specific conceptual framework prior to fieldwork. On completion of fieldwork, it becomes important to relate findings to existing theories and other published research. The overarching importance granted by interviewed stakeholders to interactions and social factors can be easily understood if one subscribes to social or bio-psychosocial models of disablement (Bickenbach *et al.*, 1999; Waddell, 1996). The views expressed by stakeholders on factors influencing RTW cannot be understood using a purely biomedical model of disability.

Using the terms proposed by the World Health Organization in its International Classification of Functioning, Disability and Health (WHO, 2001), occupational disability is seen by most interviewed stakeholders as a restriction in participation heavily influenced by the interaction of personal and environmental factors inside and outside the workplace.

The themes discussed by interviewed stakeholders are strikingly similar to those described by workers and care providers involved in a managed care experiment in Washington State described by Mootz and colleagues in Chapter 9 of this volume, and by others (Brines *et al.*, 1999; Pergola and Graham, 1999). The quality assessment conceptual framework used by the Washington State researchers had Structure, Process and Outcome as the main domains of interest (Salazar and Graham, 1999). Structure broadly corresponds to the social, economic and regulatory context factors identified in the present study, while Process relates to the stakeholder interactions described in this chapter. Workplace factors were not as salient in the Washington State study, perhaps because employers and supervisors were not specifically targeted (Salazar and Graham, 1999).

5.4.2 Implications for Practice

Perhaps the most important lesson for RTW practitioners from the Work-Ready project is to "look at the big picture without losing sight of the details". In setting up an RTW program, or in trying to facilitate the RTW of a particular worker after a soft tissue injury, one should always consider the three levels depicted in Figure 5.1. Systematic consideration of the elements depicted in Figure 5.1 will often point out to those areas which are acting as "bottle-necks" in the situation at hand, and will suggest possible courses of action (Shrey, 1996; Loisel *et al.*, 2001).

The second important lesson is that human factors and interactions are crucial. Practitioners taking on the role of facilitators of RTW after a soft tissue injury ought to avoid finger-pointing, and concentrate on promoting constructive interactions between the stakeholders involved. They ought to strive to improve these interactions by facilitating open, honest communication and building mutual trust and commitment. Successful RTW is a "win-win" for all concerned, a fact that must sometimes be repeatedly but diplomatically pointed out (Shrey, 1996).

When addressing the first level, i.e., the interactions within the workplace, practitioners ought to try to influence workplace stakeholders to harmonize prevention and RTW goals with production and performance goals, at all levels of the organization. In managing the case of an individual injured worker, this type of major organizational change is not likely to be considered as an immediate option. A more realistic approach may be to try to empower the worker with knowledge and skills that will allow him or her to more effectively negotiate through the complex workplace, healthcare, and disability insurance systems.

Workplaces, when embarking on a program to enhance RTW, may concentrate most of their attention on one type of solution, for example, an ergonomic approach. This allows a clear focus, and attention to concrete solutions, which can be understood and generally accepted by all concerned. The results of the Work-Ready study suggest that an individualized people-centered approach and a firm commitment to also work on the more intangible aspects of the problem, such as beliefs and attitudes, is critical to ultimate success.

Incorporation of the RTW program into comprehensive occupational health and safety programs addressing prevention is important. Besides the benefits directly derived from such prevention activities, this integration will send a strong message to everybody in the organization that the employer is genuinely concerned about employee health and welfare and will strengthen the commitment to RTW.

5.4.3 Implications for Research

One fundamental decision made in the initial phase of the Work-Ready study was to utilize qualitative research methods as the most effective way to gather and analyze perspectives of occupational health stakeholders. This decision allowed project flexibility and the ability to explore issues in-depth as they arose. Although we suspected organizational and interpersonal factors were important in RTW, it was eye-opening to see the importance that most occupational stakeholders assign to these aspects of the RTW process. Qualitative research methods allowed us to obtain more in-depth information on the complexity and dynamics of the RTW process that quantitative methods cannot capture. We strongly encourage the use of qualitative research methods, as an adjunct to more traditional quantitative methods, in future studies of the RTW process.

The great number and variety of RTW programs currently in use in the workplaces studied contrast markedly with the paucity of rigorous evaluations of their efficacy, effectiveness and costs (Baril *et al.*, 2000; Loisel *et al.*, 2001). Most programs are put in place because legislation has made it advantageous or necessary, because they are heavily promoted by private vendors in business publications and seminars, or because they are compatible with local management attitudes and workplace culture. Proof of their effectiveness, let alone their cost-effectiveness, is almost never a pre-requisite.

There is a clear need for rigorous studies to evaluate the impact and efficiency of specific RTW programs. Initially, studies could be observational, taking advantage of the "natural experimental" variation across provinces and firms, but they should pay attention to explicitly describing the social, economic, and workplace cultural context of each RTW program.

Nevertheless, observational studies are not enough. Experimental (or at least quasi-experimental) studies, i.e. comparisons of two or more competing RTW programs implemented over time across a number of comparable workplaces, are needed to establish the effectiveness of specific programs. A good example is the Sherbrooke study by Loisel *et al.*, which compared workplace and clinical components of four RTW programs in 32 randomized workplaces (Loisel *et al.*,

1997). For reasons of feasibility, and to facilitate comparisons among competing programs, studies should probably focus on workplaces within a single sector or local setting. Throughout the conduct of these experimental studies, the perceptions and opinions of all stakeholders involved should be carefully recorded, to fully judge the effects of the program, to allow comparison with other contexts, and to further our understanding of why a particular program was or was not successful.

In conclusion, our contacts with occupational health stakeholders involved in different aspects of the RTW process after soft tissue injury, in three Canadian provinces, have lead us to consider the three-level model of facilitators and barriers shown in Figure 5.1, and to offer some concrete suggestions for better practice and research in this field. We believe the model and suggestions will be useful for decision making by all of those who have a stake in facilitating recovery and decreasing disability of workers with occupational soft tissue injuries.[2]

5.5 KEY MESSAGES

- Judging from the perspective of occupational health stakeholders, three kinds of factors (listed in order of importance) influence RTW after an occupational injury.

  a) Interactions within the workplace, including elements of the workplace culture, organizational structure and work organization.

  b) Interactions outside the workplace, particularly those involving the health care system and the disability insurance system

  c) The characteristics of the social, economic and regulatory context which modulate the interactions

- RTW succeeds when the interactions listed above are characterized by open and honest communication, commitment to the RTW process and trust in the other stakeholder.

- Those involved in facilitating RTW should consider all three kinds of factors and concentrate on promoting constructive interactions among the stakeholders.

- Multi-stakeholder case-based workshops are a good way to open dialogue among local stakeholders to initiate collaborative action on RTW.

5.6 NOTES

[1]**Ontario Project Working Group**
John Frank – family physician/epidemiologist
Claire Bombardier – rheumatologist/epidemiologist

Judy Clarke – psychometrist/anthropologist
Donald Cole – occupational physician/epidemiologist
Pierre Côté – chiropractor/epidemiologist
Jaime Guzman – rheumatologist/epidemiologist
Vicki Pennick – occupational health nurse/community health

Quebec Project Working Group
Susan Stock – occupational physician/epidemiologist
Raymond Baril – anthropologist
Suzanne Deguire – sociologist
Marie-José Durand – occupational therapist/epidemiologist
Patrick Loisel – orthopaedic surgeon/epidemiologist
Michel Rossignol – occupational physician/epidemiologist

Manitoba Project Working Group
Annalee Yassi – occupational physician/epidemiologist
Juliette Cooper – occupational therapist/rehabilitation research
Margaret Friesen – occupational therapist/adult educator

[2]At the same time that the fieldwork described in this chapter evolved, the research team developed a case-centered tool-kit to help build bridges between the perspectives of the different stakeholder groups. This tool-kit was refined through a series of occupational stakeholder workshops in British Columbia, Manitoba and Ontario. Workshop evaluations suggest that case-based workshops using the Work-Ready tool-kit are an effective way of engaging occupational stakeholders in a constructive dialogue to exchange perspectives and interpretations of the available scientific evidence on the RTW after an occupational soft tissue injury. This dialogue is an essential pre-requisite for effective regional collaborative action to minimize the occupational disability associated with soft tissue injuries.

The Work-Ready tool-kit includes:

- a lay-language summary of evidence and experience on recovery and RTW after a soft tissue injury in the form of a pre-workshop booklet,

- nine revised case studies discussing relevant research evidence and experience in the context of individual worker vignettes,

- a facilitator's manual detailing the key activities necessary for effective conduction of a Work-Ready workshop,

- a set of auxiliary tools to recruit participants and evaluate the workshops.

As a whole, this tool-kit constitutes a detailed road map on how to bring occupational health stakeholders together to share and interpret available research evidence on the management of the occupational disability associated with soft tissue injuries.

5.7 REFERENCES

Baril, R., Martin, J.C., Lapointe, C. and Massicotte, P., 1994, Etude exploratoire des processus de réinsertion sociale et professsionnelle des travailleurs en réadaptation. *Etudes et Recherches*, (IRST).

Baril, R., Berthelette, D., Ross, C., Gourde, D., Massicotte, P. and Pajot, A., 2000, *Les composantes et les déterminants organisationnels des interventions de maintien du lien d'emploi en entreprises* (R-238), Montréal, Institut de recherche Robert-Sauvé en santé et sécurité du travail du Québec.

Bickenbach, J.E, Catterji, S., Badley, E.M. and Ustun, T.B., 1999, Models of disablement, universalism and the international classification of impairments, disabilities and handicaps. *Social Science & Medicine*, **48**, pp.1173-1187

Brines, J., Salazar, M.K., Graham, K.Y., Pergola, T. and Connon, C., 1999, Injured workers' perceptions of case management services. A descriptive study. *American Associacion of Occupational Health Nurses Journal*, **47**, pp, 355-364.

Clarke, J., Cole, D. and Ferrier, S., 1999, *Work-Ready Ontario: Stakeholder perspectives on return-to-work*, HEALNet Work-Ready Research Group.

Frank, J.W., Brooker, A.S., DeMaio, S.E., Kerr, M.S., Maetzel, A., Shannon, H.S., Sullivan, T.J., Norman, R.W. and Wells, R.P., 1996, Disability resulting from occupational low back pain. Part II: What do we know about secondary prevention? A review of the scientific evidence on prevention after disability begins. *Spine*, **21**, pp. 2918-2929.

Frank, J.W., Sinclair, S., Hogg-Johnson, S., Shannon, H., Bombardier, C., Beaton, D. and Cole, D., 1998, Preventing disability from work-related low-back pain. New evidence gives new hope – if we can just get all the players onside. *Canadian Medical Association Journal*, **158**, pp. 1625-1631.

Friesen, M.N., Yassi, A. and Cooper, J., 1999, *Work-Ready Manitoba. Stakeholder perspectives on return-to-work*, (HEALNet Work-Ready Research Group).

Gliner, J.A., 1994, Reviewing qualitative research: proposed criteria for fairness and rigor. *The Occupational Therapy Journal of Research*, **14**, pp. 78-90.

Hager, W.D., 1993 *Workers compensation back claims study, national council on compensation insurance*, (Boca Raton, FL).

Krefting, L.H. and Krefting, D.V., 1991, Cultural influences on performance. In *Occupational Therapy* edited by Christiansen, C. and Baum, C., (Thorofare, NJ: Slack Incorporated), pp. 99-122.

Liberty International Canada, 1995, Unfolding change: workers' compensation in Canada: a report for Canadians in five volumes. *Liberty International*, (Toronto: CorpWorld Group).

Loisel, P., Abenhaim, L., Durand, P., Esdaile, J.M., Suissa S., Gosselin, L., Simard, R., Turcotte, J. and Lemaire, J., 1997, A population-based randomized clinical trial on back pain management. *Spine*, **22**, pp. 2911-2918.

Loisel, P., Durand M-J., Berthelette, D. *et al.*, 2001, Disability prevention: the new paradigm of management of occupational back pain. *Disease Management & Health Outcomes*, in press.

Mondloch, M. and Cole, D., 1999, ECC Prognostic Modelling Group. Do injured worker's recovery expectations predict actual outcomes? *Working paper #73.* (Toronto: Institute for Work & Health).

Mondloch, M., Cole, D. and Frank, J., 2001, Does how you do depend on how you think you'll do? A systematic review of the evidence for a relation between patient's recovery expectations and outcomes. *Canadian Medical Association Journal*, **165**, pp. 174-179.

Pergola, T., Salazar, M.K., Graham, K.Y. and Brines, J., 1999, Case management services for injured workers. Providers' perspectives. *American Associacion of Occupational Health Nurses Journal*, **47**, pp. 397-404.

Salazar, M.K. and Graham, K.Y., 1999, Evaluation of a case management program. Summary and integration of findings. *American Association of Occupational Health Nurses Journal*, **47**, pp. 416-423.

Shrey, D.E., 1996, Disability management in industry: the new paradigm in injured worker rehabilitation. *Disability & Rehabilitation*, **18**, pp. 408-414.

Smircich, L., 1983, Concepts of culture and organizational analysis. *Administrative Science Quarterly*, **28**, pp. 339-358

van Tulder, M., Koes, B.W. and Bouter, L.M., 1995, A cost-of-illness study of back pain in The Netherlands. *Pain*, **62**, pp. 233-240.

Waddel, G., 1996, Low back pain: a twentieth century health care enigma. *Spine*, **21**, pp. 2820-2825.

Work-Ready Research Group. Facilitation of RTW after a soft tissue injury: Synthesizing evidence and experience, Report of *Work-Ready Phase 1* to HealNet, 1999.

World Health Organization, 2001, *International Classification of Functioning, Disability and Health*. hyperlink http://www.who.org

CHAPTER 6

Role of the Media in Disability Management

Rachelle Buchbinder, Damien Jolley and Mary Wyatt

6.1 LOW BACK PAIN IS AN ORDINARY FEATURE OF EVERYDAY LIFE

Most population-based surveys of back pain report a lifetime prevalence of 60% to 80% with a point prevalence of 15% to 30% and a one-year prevalence of 50% (Nachemson *et al.*, 2000). Back pain constitutes the second most common symptom prompting general practitioner visits after upper respiratory complaints (Bridges-Webb *et al.*, 1992; Cypress, 1983; Hart *et al.*, 1995). It is the most common cause of activity limitation in adults less than 45 years and the fourth most common in those aged 45 to 64 years (Andersson, 1997). While most episodes of acute low back pain improve, longitudinal studies suggest that back pain is often persistent and typically a recurrent condition (von Korff, 1994; van den Hoogen *et al.*, 1998; Croft *et al.*, 1998; Schiottz-Christensen *et al.*, 1999; Cherkin *et al.*, 1996). Despite this, the majority of people experiencing acute back pain are able to resume normal function and return to work quickly whether the pain has fully resolved or not (von Korff, 1994; van den Hoogen *et al.*, 1998; Schiottz-Christensen *et al.*, 1999).

6.2 MANAGEMENT OF SIMPLE ACUTE LOW BACK PAIN BASED UPON BEST AVAILABLE EVIDENCE

As Loisel and Durand note in Chapter 3, management of simple acute low back pain should consist of supportive advice to stay as active as possible and continue normal daily activities within the limits permitted by the pain. There is now good evidence that bed rest should not be recommended as a treatment for simple back pain and exercise therapy is ineffective in the acute phase (Malmivaara *et al.*, 1995; Deyo, 1996; Koes *et al.*, 1996). Analgesics, beginning with paracetamol, prescribed at regular intervals, non-steroidal anti-inflammatory drugs, and spinal manipulation may provide short-term symptomatic relief for uncomplicated acute low back pain (Deyo, 1996; Koes *et al.*, 1996). Patients who have not returned to ordinary activities and work by six weeks should be referred for an exercise program (Malvimaara *et al.*, 1995; van Tulder *et al.*, 1997a; Moffett *et al.*, 1999; Carey, 1995; Frost *et al.*, 1995). In general, radiography, imaging, and specialist referral are unnecessary in the acute phase.

Since the mid-1980s patients' attitudes and beliefs, particularly fear-avoidance beliefs, pain-coping strategies and illness behaviors have been increasingly recognized to be important predictors of outcome (Lethem *et al.*, 1983; Waddell, 1987; Waddell *et al.*, 1993; Symonds *et al.*, 1996). They are strongly linked to the transition from acute to chronic back pain disability and are more important than biomedical or biomechanical factors (Linton, 2000; Klenerman *et al.*, 1995; Burton *et al.*, 1995). The fear-avoidance model sets out to explain why the majority of people with low back pain recover spontaneously while a small minority develops chronic problems (Klenerman, 1995). The premise is that there are individual differences in response to painful stimuli that can be seen in terms of a continuum of fear of pain. On the one hand, there are "confronters" who have minimal fear of pain and are able to gradually increase their exposure to painful activities and rehabilitate themselves. "Avoiders", on the other hand, have a strong fear of pain leading them to avoid activities that may exacerbate their pain, which leads to other somatic consequences such as loss of mobility and muscular strength, which in turn reinforces avoidance behavior and the sick role (Klenerman *et al.*, 1995). The predictive value of this explanatory model has been shown for individuals with both acute and chronic low back pain, signifying the importance of psychosocial factors in both the assessment and management of low back pain (Waddell *et al.*, 1993; Klenerman *et al.*, 1995).

Efforts by employers to accommodate the low back pain sufferer at work have also been shown to be important in reducing duration of disability due to back pain (Loisel *et al.*, 1997; Rossignol *et al.*, 2000). Appropriate modification of work duties, enabling either no lost time or early return to work has been shown to reduce both the duration of back claims and the incidence of new back claims (Frank *et al.*, 1998). Newer studies of guideline-based approaches to back pain in the workplace suggest that coordinated workplace-linked care systems can achieve a reduction of 50% in time lost due to back pain at no extra cost and in some settings with significant savings (Frank *et al.*, 1998).

6.3 DISABILITY ASSOCIATED WITH LOW BACK PAIN IS A GROWING PUBLIC HEALTH CONCERN

Notwithstanding the good prognosis of acute low back pain and increasing scientific evidence for simple conservative management (van Tulder *et al.*, 1997a; Waddell *et al.*, 1997), disability or restricted function due to back pain is becoming an increasing public health concern in developed countries worldwide. In Australia, back problems are the leading specific musculoskeletal cause of health system expenditure with an estimated total cost of $A700 million in 1993-1994 (Mathers and Penn, 1999) and these costs are rising. In the Australian state of Victoria alone, back claims cost the workers' compensation scheme $A510 million in the 1999/2000 financial year (Victorian WorkCover Authority, 2001). While back pain accounts for about one-third of all workers' compensation lost-time claims, it accounts for 40% of all long-term claims and almost 50% of the total costs. Estimates of the costing for workers' compensation suggest that the costs for employers have been at least matched by similar amounts for the individual and the community (Industry Commission Report, 1995). Individuals who develop chronic back pain may become isolated and depressed and may have lengthy absences from work and/or become unemployable in the long term (Hendler, 1984). Disability is also frequently associated with family disruption, loss of self-esteem and quality of life (Philips and Jahanshahi, 1986; Zarkowska and Philips, 1986).

6.4 DISABILITY ASSOCIATED WITH LOW BACK PAIN WITHIN A SOCIAL CONTEXT

Disability due to low back pain needs to be viewed within a social context (Loeser and Sullivan, 1995; Waddell, 1996). Development of disability will be influenced by societal factors such as prevailing community views, political agendas of governing powers and the existing legislation regarding sickness absence and compensation. Social influences have also been shown to play a more important role than scientific influences in shaping behavior and medical decisions of physicians (Dixon, 1990). While management of low back pain will be shaped by the treater's knowledge, beliefs and training, it is also influenced by these environmental factors. The patient's presentation of their symptoms, and their knowledge, beliefs and expectations from the encounter will also influence management (Chew Graham and May, 1999; Little *et al.*, 1998; Owen *et al.*, 1990; Freeborn *et al.*, 1997). For example, patient satisfaction, reassurance, relationship to work and maintenance of the doctor-patient relationship have all been shown to be important psychosocial determinants of physician behavior when managing back pain. Patients' concerns about the seriousness of their condition, the explanatory models they construct to understand their symptoms, and their general views and expectations also play a major role (Kravitz *et al.*, 1997).

6.5 LACK OF A WIDELY ACCEPTED THEORY FOR EXPLAINING NON-SPECIFIC LOW BACK PAIN

Patients expect a diagnosis, and therefore most patients are not satisfied with the term "non-specific" low back pain (Borkan *et al.*, 1995). Depending upon the health professional that is seen for acute low back pain, widely disparate and conflicting explanations of the cause of symptoms may be given, reflecting lack of a consistent widely acceptable theory for explaining non-specific back pain (Cherkin *et al.*, 1994; Cedraschi *et al.*, 1998). The availability of more sophisticated investigations has done nothing to alleviate the problem. While investigations may be ordered to assuage patient and physician worry (Owen *et al.*, 1990) they can have detrimental effects. Patients may be given diagnostic labels based upon the appearance of imaging although the correspondence between symptoms and anatomical findings is poor (van Tulder *et al.*, 1997b; Wiesel *et al.*, 1984; Boden *et al.*, 1990; Jensen *et al.*, 1994). For example, roughly 40% of patients with "advanced disc degeneration" on plain radiograph do not have back pain (Roland and van Tulder, 1998). To highlight the limited predictive value of radiographic findings, Roland *et al.* suggested a set of statements such as this to be used in standard reports (Roland and van Tulder, 1998). Terms such as ruptured, torn, prolapsed or herniated disc may be misinterpreted by both health professional and patient, leading to fear, inappropriate behavior and worsening disability.

6.6 SATISFACTION WITH MEDICAL CARE LINKED TO PROVISION OF INFORMATION

The evidence suggests patient satisfaction with care to be most highly correlated with provision of information (Skelton *et al.*, 1996; von Korff *et al.*, 1994; Carey *et al.*, 1995; Deyo and Diehl, 1986). In a qualitative review of back pain management, Skelton *et al.* found that only 22 of 55 patients reported satisfaction with levels of primary care (Skelton *et al.*, 1996). The main reported issues of concern were communication and thoroughness, and 21 patients reported dissatisfaction with the practitioners' ability to provide an adequate explanation about the problem. Von Korff *et al.* found that patient satisfaction with primary care management was positively correlated with emphasis on provision of information and advice on self-management of back pain, versus the traditional approach of prescribing analgesics and bed rest (Von Korff *et al.*, 1994). Carey *et al.* reviewed consecutive patients attending medical and chiropractic practices and found that less than 50% of patients were satisfied with the information provided, the treatment of their back problem, and the overall result of treatment (Carey *et al.*, 1995). Deyo *et al.* found patients with back pain expressed a need for more and better-quality information about their condition (Deyo and Diehl, 1986). Unmet needs were also associated with poorer compliance and a desire for more investigation or evaluation. These studies all suggest that primary care physicians and others may lack knowledge and/or skills and/or confidence in imparting information about back pain.

6.7 IMPROVED OUTCOME ASSOCIATED WITH PROVISION OF INFORMATION AND ADVICE ADDRESSING PSYCHOSOCIAL FACTORS

A variety of studies have now shown that an informative approach designed to promote a positive approach to low back pain and address fear-avoidance beliefs and poor coping strategies improves outcomes (Roland and Dixon, 1989; Symonds *et al.*, 1995; Burton *et al.*, 1999). Roland *et al.* found than an educational pamphlet about low back trouble reduced consultation rates in a primary care clinic (Roland and Dixon, 1989). Provision of positive messages designed to improve back beliefs and reduce fear was also shown to reduce self-reported disability in those presenting with low back pain in general practice (Burton *et al.*, 1999). In an industrial setting, Symonds *et al.* demonstrated that distribution of an educational psychosocial pamphlet designed to foster positive beliefs and attitudes successfully reduced absenteeism (Symonds *et al.*, 1995). They observed that the reduction in extended work absence occurred in those with and without pre-existing back complaints (Symonds *et al.*, 1995). The benefits of these approaches may far outweigh those derived from specific traditional treatments (Indahl *et al.*, 1995).

Deyo *et al.* also showed that an educational intervention designed to modify patient expectation can reduce inappropriate imaging and costs, without compromising symptom resolution, functional improvement, satisfaction or detection of serious pathology (Deyo *et al.*, 1987).

6.8 TRANSLATING RESEARCH FINDINGS INTO CLINICAL PRACTICE

Attempts to implement research findings into practice and change physician behavior have met with limited success (Bero *et al.*, 1998; Wensig *et al.*, 1998). While some studies have shown that general practitioners manage back pain appropriately (Maetzel *et al.*, 2000), others have found that traditional approaches such as bed rest and early imaging and physiotherapy referral continue to prevail (Freeborn *et al.*, 1997; Carey and Garrett, 1996; Cherkin *et al.*, 1995; Little *et al.*, 1996; Schroth *et al.*, 1992; Elam *et al.*, 1995).

Passive dissemination of information is generally unsuccessful in influencing behavior in general practice (Bero *et al.*, 1998), whereas combinations of information transfer, learning through social influence or management support and reminders or feedback seem to be effective (Wensig *et al.*, 1998). A coordinated system of care, which accommodates physician and patient preferences, meets patient needs for adequate explanation and participation in decision making and establishes an ongoing relationship with a caring provider has been shown to effect behavior change and improve outcomes (Rossignol *et al.*, 2000).

Changing physician behavior is very complex – it involves enhancing knowledge, changing attitudes and beliefs, providing necessary skills and resources, and providing support and feedback (van Tulder *et al.*, 1997a). There are many potential sources of variation in managing an individual presenting with acute low back pain. Modifying one source of variation without concomitantly considering the effect (or lack of effect) on other factors may lead to failure to effect a change in outcome. For example, educational updates and clinical guidelines about back pain may improve physician knowledge but may not effect a change in behavior (and final outcome) if patient expectation or the doctor's beliefs about patient expectation or environmental influences remain the same. Management of low back pain commonly involves other physicians including orthopaedic surgeons, rheumatologists, rehabilitation physicians and radiologists; and other health professionals including physiotherapists and chiropractors yet most research on changing management has been directed towards primary care. It is not known whether the issues regarding changing behavior will be equally applicable to other groups.

6.9 NEED TO SHIFT SOCIETAL VIEWS ABOUT BACK PAIN

While the bio-psychosocial model of back pain proposed by Waddell (Waddell, 1987) and others in the mid-1980s is now well accepted among back pain experts, there is abundant evidence that this fundamental shift from the traditional medical model of back pain has yet to be universally adopted. Recent qualitative studies have highlighted the current disparity between the low back pain patient's physical model of pain causation and the doctors' psychosocial model (Chew-Graham and May, 1999). Doctors feel ill-equipped to challenge the patient's model without damaging the doctor-patient relationship and in any case may be unable to influence the social factors that shape the patient's presentation (Chew and May, 1997). Patients' views and beliefs about the cause and prognosis of their back pain and about the most effective treatments may not correspond to the care proposed by the evidence (van Tulder *et al.*, 1997a). With increasing emphasis on patient participation and shared decision making, patients' views and beliefs may influence the process and outcome of consultations. As previously suggested by Deyo, it may be that the public as well as the medical profession need to be re-educated (Deyo, 1996). If re-education can change population attitudes and beliefs and give rise to a concomitant alteration in patient expectation and physician behavior, this may be effective in stemming or reversing the rising epidemic of low back pain disability (Buchbinder *et al.*, 2001a, b, c).

6.10 THE VICTORIAN WORKCOVER AUTHORITY MASS MEDIA BACK CAMPAIGN

A ground-breaking new approach to the prevention and management of low back pain disability was the three-year population-based mass media campaign initiated by the Victorian WorkCover Authority (VWA) in the state of Victoria, Australia, in 1997. Entitled "Back Pain: Don't Take It Lying Down", the campaign aimed to reduce back pain related disability by challenging the traditional attitudes towards back problems and shifting beliefs in line with modern philosophy.

The messages were simple and in line with current evidence: back pain is not a serious medical problem; disability can be improved and even prevented by positive attitudes; treatment should consist of continuing to perform usual activities, not resting for prolonged periods, exercising and remaining at work (Burton *et al.*, 1999). It counseled individuals with low back pain, their doctors and employers to avoid excessive medicalization of the problem, unnecessary diagnostic testing and treatment. It tackled fear-avoidance beliefs and promoted gradual exposure to painful activities emphasizing that it was unlikely to be harmful. These messages were based upon those delineated in *The Back Book,* an evidence-based patient educational booklet produced in the UK by a multidisciplinary team of authors (Roland *et al.*, 1996). To ensure the campaign did not discourage patients with serious pathology from seeking early medical assessment, advertisements highlighting "red flag" symptoms were also shown.

The campaign included electronic, print and outdoor media and targeted both the community and treating doctors. The major medium was television commercials aired in prime time slots commencing in September 1997. The intensity of the campaign varied with a concentrated campaign for three months initially followed by a low-key maintenance campaign until September 1998. A further three-month concentrated television campaign commenced in September 1999 followed by a low-key maintenance campaign until February 2000. The messages of each of the television commercials are shown in Table 6.1.

Table 6.1 Individual television commercial messages (continued overleaf).

Messages	Delivered by
Exercise, rest, work, shift responsibility of control to individual with low back pain	
• back pain...don't take it lying down.	Billboards
• Bed rest often makes it worse, stay active and at work if you can.	International back expert/ orthopaedic surgeon
• Revolution in backache, stay active and at work if you possibly can.	
• No magic fix. Get it moving, get it working and get it fit. It is up to you.	Local comic "personality"
• More than a few days bed rest and the brain goes soft. Better to return to work, just take it easy.	General practitioner
• Not one quick fix. Paracetamol or aspirin, heat or ice, manipulation can be helpful in the first six weeks, but in the end it's up to you, get moving.	General practitioner
• Movement helps ease the pain and helps healing, you don't want to let your life pass you by.	General practitioner
• Rest is bad for backs; get moving early.	
• The spine is difficult to damage. If it hurts it is likely to settle quickly.	Orthopaedic surgeon
• Walk away from a negative attitude and back problem, back into job and into life.	Actor depicting back pain sufferer
• Straining your back at work is similar to a sports injury and the treatment should be the same. Check it out; keep it active.	Sports physician
• Exercise helps my pain. Give exercise a go.	Well known Australian ex-cricketer
• Keep those strains moving, lifting equipment at work.	
• I've had back pain for years. Hurt doesn't mean harm; keep active.	Well known Australian-rules footballer
• Need a little treatment, a little advice and exercise. With a little help nearly everyone can work.	Osteopath
• Experts around the world agree, exercise is good for back pain, don't let it be too late for you.	Minister for Health

• It is good to exercise and not worry too much about back pain, but if you have constant and unremitting pain, or numbness below the waist see your doctor straight away.	General practitioner
• Treat back pain like a simple cold, don't put your life on hold, don't let backache cripple.	International back expert
• The spine is like a mask. Strengthen muscles through regular exercise.	Physiotherapist
• You can do a lot for your back, exercise, stay at work if you can, give it a go.	Chiropractor
• Lives have been ruined by heavy analgesics or surgery. Keep moving, keep active. For simple back pain the best treater can be you.	International back expert, general physician

Investigations may not be helpful in finding a cause for the pain

• Do what doctors do. Stay active, stay at work, don't rush off for an X-ray.	International back expert/ orthopaedic surgeon
• X-rays don't show pain.	Rehabilitation specialist
• Disc bulges and deterioration are normal X-ray findings.	International back expert, general physician

Spinal surgery

• Before spinal surgery – get a second expert opinion.	Spinal surgeon
• I've had back trouble since the age of 15. Surgery can't guarantee a result. Exercise is safer and a surer bet than an operation.	Well known Australian ex-cricketer
• Before you take advice that might change your life, get another opinion.	Rehabilitation specialist

Positive messages to employers about keeping employees at work (included later in the campaign – September 1999)

• Keeping injured workers on the job is better for them.	Actor
• Rehabilitation helps people get back to work when employers haven't kept them on, give employees the chance to get better at work.	Rheumatologist and rehabilitation physician
• Simple changes in the workplace can allow workers with back pain to return to their job before they have fully recovered. The same changes can prevent injuries.	Occupational physician

Commercials included recognized international and national medical experts in orthopaedic surgery, rehabilitation, rheumatology, general practice, physiotherapy, chiropractic therapy, sports and occupational medicine. There were also advertisements by well-known sporting and local television personalities who had successfully managed their own back pain. All advertisements concluded with endorsements by relevant national or state professional medical bodies.

The television campaign was supported by radio, outdoor and print advertisements, posters, seminars, visits by personalities to workplaces, publicity articles and publications. *The Back Book* (Roland *et al.*, 1996) was sent to all treating practitioners in Victoria to be given to patients suffering from back pain, and in latter stages of the campaign to insurers to be passed on to claimants. Guidelines for the management of employees with compensable low back pain were simultaneously developed over a three-year period with consultation from a wide variety of treating practitioners (Victorian WorkCover Authority, 1996).

6.11 DID IT WORK?

We undertook an independent three-part evaluation of the campaign and our results suggest that there has been widespread adoption of these messages (Buchbinder *et al.*, 2001a; Buchbinder *et al.*, 2001b,c).

6.11.1 Population Beliefs

We measured the effect of the campaign on population beliefs by conducting telephone surveys of three separate cross-sectional random samples of the employed population prior to the campaign and two and two and a half years after campaign onset. The telephone surveys were conducted in Victoria and New South Wales (NSW), an adjacent state where we knew no state-wide public health campaign for back pain was to take place for the duration of our study. Both states have similar demographic characteristics and worker's compensation systems.

There were 4,730 surveys completed, with equal numbers in each state, and a ratio of 2:1:1 across the three time periods. Demographic characteristics and previous experience of back pain were similar at each time point both within and between states. At baseline about half of the respondents in both states had been aware of back pain advertising in the previous year (Victoria 47.1%, NSW 51.5%). While this did not change over time in NSW (48.5% and 48.0% for Survey 2 and 3 respectively), there was a significant increase in awareness of back pain advertising in Victoria (73.9% and 85.5% for Survey 2 and 3 respectively) ($p < 0.001$). This was accompanied by a self-reported change in beliefs about back pain as a consequence of advertising (23.1%, 38.8%, 48.0% at Surveys 1-3 respectively) ($p < 0.001$) (Buchbinder *et al.*, 2001).

The primary measure of beliefs was the Back Beliefs Questionnaire (BBQ), a self-administered questionnaire designed to measure beliefs about the inevitable consequences of future life with low back trouble (Symonds *et al.*, 1995). A higher score indicates a more positive belief about low back trouble, suggesting better ability to cope with low back pain. At baseline, mean BBQ scores were similar in both states (Table 6.2). There were significant improvements in mean BBQ score in Victoria between successive surveys, whereas the mean BBQ score was unchanged in NSW across time (Buchbinder *et al.*, 2001).

At baseline the distribution of BBQ scores was almost identical in both states. At Survey 2 BBQ scores were generally higher in Victoria compared with NSW across the range of BBQ scores with a further shift towards higher scores in Victoria at Survey 3. As well as observing a uniform shift in BBQ scores in Victoria irrespective of baseline score, improvements in the mean BBQ score were also observed in Victoria irrespective of baseline demographic or other clinical characteristics (Figure 6.1). Statistically significant improvements in mean BBQ scores in Victoria were seen between successive surveys irrespective of age, sex, education level, employment status, type of work (manual or not), income, previous back pain experience, whether respondents reported awareness of back pain advertising or not, and irrespective of whether attitudes were reported to have changed as a result of seeing an advertising campaign (Buchbinder *et al.*, 2001a; Buchbinder *et al.*, 2001b,c).

Table 6.2 Back Beliefs Questionnaire (BBQ) and Fear-Avoidance Beliefs Questionnaire (FABQ) – mean scores (95% CI) and difference in mean score from Survey 1 (95% CI) by state (Victoria, NSW) and survey wave (1 – August 1997, 2 – August 1999, 3 – February 2000)

State	Survey	*n*	Mean score (95% CI)	Difference in mean score from Survey 1 (95% CI)	P-value
BBQ (possible score 9 - 45)					
Victoria	1	1,185	26.5 (26.1 - 26.8)		
	2	590	28.4 (27.9 - 28.8)	1.9 (1.3 - 2.5)	0.000
	3	592	29.7 (29.2 - 30.3)	3.2 (2.6 - 3.9)	0.000
NSW	1	1,185	26.3 (25.9 - 26.6)		
	2	590	26.2 (25.7 - 26.7)	-0.04 (-0.7 - 0.6)	0.9
	3	588	26.3 (25.7 - 26.8)	0.02 (-0.6 - 0.7)	1.0
FABphys (possible score 0 - 24)					
Victoria	1	640	14.0 (13.6 - 14.4)		
	2	343	12.5 (11.9 - 13.1)	-1.5 -2.2 to -0.8	0.000
	3	307	11.6 (11.0 - 12.2)	-2.4 -3.1 to -1.6	0.000
NSW	1	645	13.3 (12.9 - 13.8)		
	2	317	13.6 (13.1 - 14.2)	0.3 (-0.4 - 1.0)	0.411
	3	274	12.7 (12.0 - 13.5)	-0.6 (-1.4 - 0.2)	0.165
Modified FABwork (possible score 0 - 36)					
Victoria	1	640	13.5 (12.7 - 14.2)		
	2	343	14.5 (13.5 - 15.5)	1.1 (-0.2 - 2.3)	0.105
	3	307	12.5 (11.3 - 13.6)	-1.0 (-2.3 - 0.3)	0.146
NSW	1	645	13.7 (12.9 - 14.4)		
	2	317	13.6 (12.5 - 14.6)	-0.1 (-1.4 - 1.2)	0.891
	3	274	12.1 (11.0 - 13.2)	-1.6 (-2.9 - -0.2)	0.020

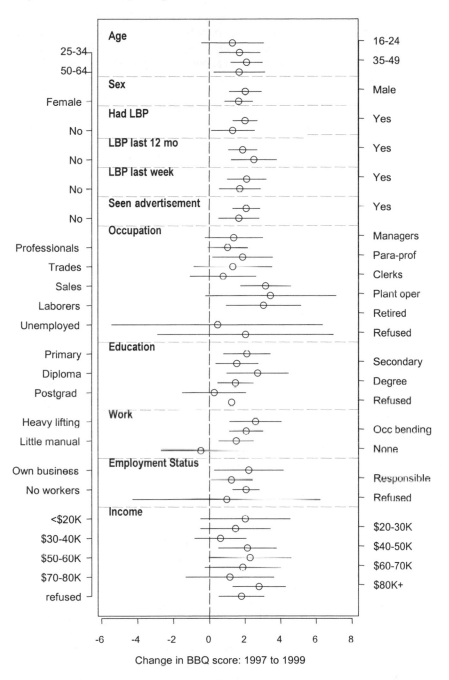

Figure 6.1 Average change (95% Confidence Interval) in score of Back Beliefs Questionnaire (BBQ) between Survey 1 and 3 in Victoria according to demographic and other characteristics.

A modified Fear Avoidance Beliefs Questionnaire (FABQ) was also administered to respondents who reported back pain in the past year. This instrument measures an individual's beliefs about physical or work activity being the cause of their trouble, and their fears about the dangers of such activities when they have an episode of low back trouble (Waddell *et al.*, 1993). It has two subscales: physical activity (FABphys) and work activity (FABwork). For both components, a lower score indicates less fear-avoidance beliefs. Significant changes also occurred in mean FABphys in Victoria between successive surveys with a mean improvement of −1.5 and −2.4 between baseline and Surveys 2 and 3 respectively, with no change in FABphys in NSW (Table 6.2). The modified FABwork did not change in either state between baseline and Survey 2 but changed in both states between baseline and Survey 3 (p=0.02 in NSW) (Buchbinder *et al.*, 2001a).

6.11.2 General Practitioner Surveys

A mailed survey of a random sample of general practitioners before and two and a half years after campaign commencement was also performed. The design was similar to the population surveys in that a non-randomized, non-equivalent before and after design was used with NSW acting as control group. The questionnaire was modified from one developed in Ontario, Canada (Bombardier *et al.*, 1995) and included a set of questions aimed at eliciting knowledge about the management of acute low back pain; attitudes towards these patients; and attitudes towards management guidelines. Questions were phrased as statements and responses on a five-point Likert scale that varied from strongly agree to strongly disagree. The respondents' likely approach to the management of acute and subacute low back pain was also elicited, based upon presentation of two hypothetical scenarios.

There were 2,556 surveys completed. As for the population surveys, the number of doctors who were aware of a back pain media campaign rose significantly in Victoria between successive surveys (from 15.5% at Survey 1 to 89.3% at Survey 2 (p< 0.001)).

Over time, Victorian doctors were 3.6 (2.4-5.6) times as likely to know that patients with low back pain need not wait to be almost pain free to return to work. They were also 2.9 (1.6-5.2) times as likely to know that patients with acute low back pain should not be prescribed complete bed rest until the pain goes away and 1.6 (1.2-2.3) times as likely to know X-rays of the lumbar spine are not useful in the work up of acute low back pain. Victorian doctors also indicated they would be less likely to order tests for low back pain either because patients expect them to (Odds Ratio (OR)=1.6 (95% CI 1.1-2.3)) or to conform with the normal practice patterns of their peer group (OR=1.5 (95% CI 1.1-2.0)) (Buchbinder *et al.*, 2001).

Over time, Victorian general practitioners were 2.51 times as likely not to order tests for acute low back pain and 0.40 times as likely to order lumbosacral X-rays. They were also 0.48 times as likely to prescribe bed rest and 1.65 times as likely to advise work modification. Similar changes were seen for subacute low back pain (Table 6.3) (Buchbinder *et al.*, 2001).

Table 6.3 Relative decrease in population disability rate for different combinations of relative risk of "high" and "medium" risk groups, assuming population changes as in Table 6.4.

Relative risk in "high" group	Relative risk in "medium" group		
	1.0	1.5	2.0
2.0	3%	8%	11%
5.0	10%	13%	16%
10.0	19%	20%	22%
15.0	25%	25%	26%
20.0	29%	29%	30%

6.11.3 Victorian WorkCover Authority Claims Database

Finally, a descriptive analysis of the VWA claims database was performed to determine the impact of the campaign upon the incidence, duration and costs of back pain claims. The number of back claims in comparison to non-back claims declined by greater than 15% over the duration of the study (Buchbinder *et al.*, 2001a). This was accompanied by a decline in the rate of days compensated for back claims steeper than that seen for non-back claims and a 20% reduction in medical payments per back claim seen over the duration of the campaign (Buchbinder *et al.*, 2001b,c).

6.12 WHAT FACTORS MAY HAVE BEEN IMPORTANT IN THE SUCCESS OF THE VICTORIAN WORKCOVER AUTHORITY CAMPAIGN?

The success of the campaign has been attributed to many factors. Firstly the evidence-based content and unambiguous direct approach first pioneered by the authors of *The Back Book* (Roland *et al.*, 1996) needs to be acknowledged. The messages were conveyed in simple language and were delivered by well-recognized local, national and international individuals. The use of positive role models such as sporting or entertainment stars is a recognized powerful attribute of mass media campaigns (Redman *et al.*, 1990). In addition, virtually every professional body with a stake in back pain in Australia supported the campaign and had input into the wording and content of the advertisements.

6.13 RATIONALE FOR A PUBLIC HEALTH POPULATION-BASED APPROACH

We demonstrated that a public policy initiative directed towards changing societal views towards back pain can be highly successful. The media campaign sought to promote positive beliefs about back pain, encourage self-coping strategies and continued activity and reduce negative beliefs about the inevitable consequences of back pain. There are compelling arguments for this approach (Buchbinder *et al.*, 2001a). Firstly, disability related to low back pain can be considered a public health issue because it affects a substantial proportion of the community and involves the use of substantial common resources. Individual clinical treatments have had limited impact upon long-term outcomes for back pain (Deyo, 1996; van der Weide *et al.*, 1997). There is only modest evidence that treatment significantly helps symptoms (van Tulder *et al.*, 1997a), and a substantial amount of evidence about the adverse effects of rest, surgery, medication, and focusing on the problem itself (Malmivaara *et al.*, 1995; Deyo, 1996; Indahl *et al.*, 1995).

Second, as described above, psychosocial interventions have been shown to reduce disability and work absence associated with low back trouble (Symonds *et al.*, 1995). These interventions aim to reduce the risk of chronicity by changing beliefs with concomitant behavioral modification leading to adoption of an early active management strategy (Burton *et al.*, 1996). The data suggest that informative interventions could even be of more value when initiated early, even prior to the onset of symptoms (Symonds *et al.*, 1995).

Third, predictive models of low back pain are not presently able to identify those at risk of disability. By targeting the entire population, a public health approach will reach those hard-to-identify high-risk groups. There is also much evidence that a population strategy of universal change has greater overall effect than targeted high-risk strategies (see below).

Fourth, the population approach may be an effective way of modifying doctor and other health professional behavior, both through direct influences as well as through the change in attitudes of their patients. There is good evidence that doctors strive to please their patients, and that they are open to patients' requests to try treatments (Avorn *et al.*, 1982). The "direct-to-consumer" approach is recognized as a powerful force in pharmaceutical marketing, and it is likely that similarly effective results can be achieved by public health media campaigns as by drug company marketing schemes. Finally, employer attitudes and beliefs and workplace philosophies may also be influenced through similar means.

6.14 BENEFITS OF A POPULATION-BASED STRATEGY

The choice of a mass-media publicity campaign to communicate a health promotion message is predicated on a "population strategy" of prevention. This attempts to control the determinants of incidence in a population by shifting the whole distribution of exposure, rather than targeting a smaller, high-risk subgroup of the population (Rose, 1992).

The rationale behind this population strategy can be demonstrated using a diagram of the distribution of the BBQ in the Victorian population both before and after the campaign (Figure 6.2). We have shown that the BBQ score distribution was shifted upwards by about 3.0 units during the period of the mass media campaign, and this is reflected in Figure 6.2 by a solid bell-shaped curve (after the campaign) falling to the right of the broken-line curve (before the campaign).

Before the campaign, the lowest 5% of the population had a BBQ score less than 16, and were regarded as the "high risk" group. We might label others with BBQ scores less than the population mean (27) as "medium risk". These groups are labelled using hatched areas in Figure 6.2. After the campaign, the entire distribution has been shifted to the right, and the corresponding proportions of "high risk" and "medium risk" groups, shown now as shaded areas in Figure 6.2, are smaller.

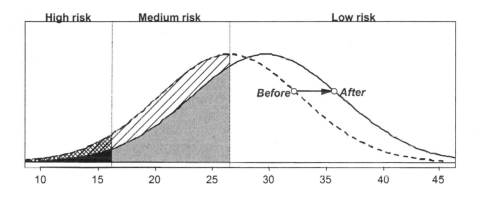

Figure 6.2 Population distribution of BBQ in Victoria before mass media campaign (broken line) and after (solid line). Proportions of the population at high and medium risk are shown as hatched areas (before) or shaded areas (after).

Table 6.4 shows the proportions of the population in each of these risk groups, both before and after the campaign. The 5% at "high risk" has dropped to only 2%, a relative decrease of 60%. The "medium risk" group decreased by more than 30%, while the "low risk" group (with BBQ score ≥ 27) increased by 33%.

Table 6.4 Relative change in proportion of the Victorian population at risk of back pain disability, before and after mass media intervention campaign.

Definition	Proportion of population		Relative change in proportion
	Before campaign	*After campaign*	
BBQ < 16	5%	2%	60% decrease
16 ≤ BBQ < 27	45%	31%	31% decrease
BBQ ≥ 27	50%	67%	33% increase

The population strategy, therefore, has an effect of reducing not only the high-risk minority of potential disabilities, but also substantially reducing the medium to low risk groups. Rose's "prevention paradox" reminds us that it is usually the medium to low risk groups in a population that account for the majority of the illnesses eventually seen (Rose, 1992).

Of course, the actual population impact of the population strategy depends on the absolute risks associated with these labelled risk groups. We have not been able to estimate these risks in this evaluation study. However, it is possible to infer the overall impact on the population disability rate. Table 6.3 shows the relative decrease in population disability rate, using a range of relative risks for disability associated with each of the "high" and "medium" risk categories as described above. For example, if the "high" risk group has a ten-fold risk of back pain disability (relative to the majority at "low risk"), and the "medium" risk group has 50% increased risk (Relative Risk=1.5), then the overall impact of the campaign will be a 20% decrease in population disability rate.

6.15 MEDIA CAMPAIGNS AND THEIR PROPER EVALUATION

The mass media plays a potentially vital role in delivering health messages to the public. It offers possible means for reaching larger numbers of people at less expense than that associated with face-to-face services. Population surveys have shown that the media is a leading source of information about important health issues (Redman *et al.*, 1990) and the popular press has also been shown to amplify the transmission of medical information to the scientific community (Phillips *et al.*, 1991). Media advocacy is an established health promotion strategy, partly influenced by the World Health Organization's Ottawa Charter for Health Promotion (Grilli *et al.*, 2001). It has been used to influence behavior of patients and health professionals, and promote effective and efficient use of health services (Grilli *et al.*, 2001). For example it has been used to alter a wide variety of health-related behaviors such as smoking (Sly *et al.*, 2001), sunlight exposure (Hill *et al.*, 1993) and physical activity (Owen *et al.*, 1995). It has also been used to promote use of healthcare interventions such as preventive asthma therapy, cancer and HIV screening, immunisation programs, and emergency services for suspected myocardial infarction (Grilli *et al.*, 2001). Media campaigns have been particularly successful in Australia in altering health-related behaviors such as smoking (Pierce *et al.*, 1990) and sunlight exposure (Hill *et al.*, 1993). In the case of sunlight exposure, the data indicate that suntans are no longer as popular as they were and people are reducing their sunlight exposure by a variety of methods with a concomitant clear reduction in sunburn, skin cancer incidence and mortality rates (Marks, 1999).

Media campaigns are also able to reach groups who are difficult to access through traditional medical delivery (Redman *et al.*, 1990). We demonstrated that the VWA campaign reached a reasonable representative sample of the general population and beliefs shifted irrespective of baseline demographic characteristics (Buchbinder *et al.*, 1997a). Another advantage of mass media campaigns over more targeted interventions is that they have the potential to modify the knowledge or attitudes of a large proportion of the community simultaneously, thereby providing social support for behavioral change (Redman *et al.*, 1990). The presence of social support may help to maintain behavioral change over time.

Based upon our current knowledge, a mass media campaign designed to influence population attitudes and beliefs about back pain and achieve a sustained change in patient and health professional behavior seems a highly appropriate health promotion strategy to adopt. It provides consumers with up to date and accurate information about back pain and its management, and gives them the confidence to deal with back pain, empower health professionals to manage back pain in a more appropriate fashion, and persuade employers to keep back pain sufferers at work.

With the current emphasis on consumerism in the delivery of healthcare, it is important to recognize that in the area of back pain, other competing interests in the area of back pain already use the mass media to convey their messages. This was illustrated in our studies by the fact that about 50% of the general population and 15-25% of general practitioners who responded to our surveys were aware of advertising campaigns about back pain prior to the onset of the VWA media campaign in Victoria. Not only is it imperative that the reporting of health-related issues in the lay media correctly represent the best available knowledge on the effectiveness of healthcare interventions, it should also discourage use of interventions of unproven effectiveness. To succeed, mass media campaigns need to be up to date, clever and attract the attention of the public.

While consistently positive effects of mass media campaigns have been observed in the published literature, systematic reviews have highlighted the methodological limitations of much of the primary research in this area (Redman *et al.*, 1990; Grilli *et al.*, 2001). Refinement of methods for evaluating the effect of media campaigns is needed including validating outcome measures and obtaining representative samples and adequate control groups. Further research is also needed to explore the characteristics of a successful media campaign including identifying the characteristics of the setting and messages that influence the effectiveness of the campaign; whether there is an equivalent effect on consumers and health professionals; the pattern and rate of uptake of the messages and change in behavior; duration of any observed effects; and cost-effectiveness of the use of mass media.

Several models exist for health-related behavioral change, but one of the more popular is the transtheoretical model of Prochaska and DiClementi (1982). Six stages of change for individuals to alter their health behavior are identified in this model, with contemplation and attitudinal change preceding actual behavioral modifications by at least six months. Permanent changes in behavior (the "termination") may not occur until a further six months or even longer after the "action" phase. Social approval also serves to reinforce attitudinal and behavioral changes. These issues need to be considered when timing evaluations of media campaigns.

6.16 IMPLICATIONS FOR HEALTH POLICY

It is hoped that the demonstration of effectiveness of the back pain media campaign will have important ramifications for further policy development in this area and innovative approaches to other health problems. However, there do not appear to be any plans to continue the campaign in Victoria. The focus of a new Victorian WorkCover Authority media campaign, implemented with the change in state government, is on introduction of ergonomic interventions in the workplace, although there is limited evidence to support the effectiveness of this approach. There are other legitimate influences on policy-making including social, electoral, ethical, cultural and economical issues (Black, 2001). Whether the results of research studies have an impact upon policy-making will also depend upon whether it is compatible with policy-makers' interests and ideologies.

It remains to be seen what the long-term impact of the VWA back campaign will be. It is likely that some ongoing reminders are necessary to maintain the changes that have occurred. Our results are now encouraging other countries, such as Sweden, the Netherlands, and Canada to consider adopting similar approaches. It will be interesting to see whether the success of the Victorian campaign can be replicated in other industrialized societies with other healthcare and disability systems. It will be important to bear in mind that while a particular intervention may have a positive impact at a local clinical level, this does not imply that large-scale implementation will be successful.

6.17 KEY MESSAGES

- A population-based media campaign that provided explicit advice about back pain:

    a) improved population beliefs about back pain
    b) improved knowledge and attitudes of general practitioners and appeared to influence their management of back pain
    c) reduced disability and workers' compensation costs related to back pain.

- The mass media is an effective way of delivering heath messages to the public.

- Shifting societal views about back pain through media campaigns may be a highly effective primary preventive strategy for reducing back pain disability.

6.18 REFERENCES

Andersson, G., 1997, *The epidemiology of spinal disorders, in the adult spine: principles and practice,* edited by Frymoyer. J.W. (Raven Press: New York), pp. 93-141.

Avorn, J., Chen, M. and Hartley, R., 1982, Scientific versus commercial sources of influence on the prescribing behavior of physicians. *American Journal of Medicine*, **73**, pp. 4-8.

Bero, L.A., Grilli, R., Grimshaw, J.M., Harvey, E., Oxman, A.D. and Thomson, M.A., 1998, Closing the gap between research and practice: an overview of systematic reviews of interventions to promote the implementation of research findings. *British Medical Journal*, **317**, pp. 465-468.

Black, N., 2001, Evidence-based policy; proceed with care. *British Medical Journal,* **323**, pp. 275-279.

Boden, S., Davis, D.O., Dina, T.S., Patronas, N.J. and Wiesel, S.W., 1990, Abnormal magnetic-resonance scans of the lumbar spine in asymptomatic subjects. *The Journal of Bone and Joint Surgery*, **72A**, pp. 403-408.

Bombardier, C., Jansz, G. and Maetzel, A., 1995, Primary care physicians' knowledge, confidence and attitude in the management of acute low back pain. *Arthritis and Rheumatism*, **38** (suppl), p. S385.

Borkan, J., Reis, S., Hermoni, D. and Biderman, A., 1995, Talking about the pain: a patient-centered study of low back pain in primary care. *Social Science and Medicine*, **40**, pp. 977-988.

Bridges-Webb, C., Britt, H., Miles, D.A., Neary, S., Charles, J. and Traynor, V., 1992, Morbidity and treatment in general practice in Australia 1990-1991. *Medical Journal of Australia*, **157** (suppl), pp. S1-S56.

Buchbinder, R., Jolley, D. and Wyatt, M., 2001a, 2001 Volvo award winner in clinical studies: effects of a media campaign on back pain beliefs and its potential influence on management of low back pain in general practice. *Spine*, **26**, pp. 2535-2542.

Buchbinder, R., Jolley, D. and Wyatt, M., 2001b, Breaking the back of back pain. *Medical Journal of Australia*.

Buchbinder, R., Jolley, D. and Wyatt, M., 2001c, Population based intervention to change back pain beliefs and disability: three part evaluation. *British Medical Journal*, **322**, pp. 1516-1520.

Burton, A., Waddell, G., Tillotson, K.M. and Summerton, N., 1999, Information and advice to patients with back pain can have a positive effect. A randomised controlled trial of a novel educational booklet in primary care. *Spine*, **24**, pp. 1-8.

Burton, A., Waddell, G., Burtt, R. and Blair, S., 1996, Patient educational material in the management of low back pain in primary care. *Bulletin – Hospital for Joint Diseases*, **55**, pp. 138-141.

Burton, A.K., Tillotson, K.M., Main, C.J. and Hollis, S., 1995, Psychosocial predictors of outcome in acute and subchronic low back trouble. *Spine*, **20**, pp. 722-728.

Carey, T., 1995, A supervised progressive fitness programme reduced pain and disability in chronic back pain. *ACP Journal Club*, **123**, p. 7.

Carey, T.S. and Garrett, J., 1996, Patterns of ordering diagnostic tests for patients with acute low back pain. The North Carolina Back Pain Project. *Annals of Internal Medicine*, **125**, pp. 807-814.

Carey, T.S., Garrett, J., Jackman, A., McLaughlin, C., Fryer, J. and Smucker, D.R., 1995, The outcomes and costs of care for acute low back pain among patients seen by primary care practitioners, chiropractors, and orthopedic surgeons. The North Carolina Back Pain Project. *New England Journal of Medicine*, **333**, pp. 913-917.

Cedraschi, C., Nordin, M., Nachemson, A.L. and Vischer, T.K., 1998, Health care providers should use a common language in relation to low back pain patients. *Baillieres Clinical Rheumatology*, **12**, pp. 1-15.

Cherkin, D.C., Deyo, R.A., Wheeler, K. and Ciol M.A., 1994, Physician variation in diagnostic testing for low back pain. Who you see is what you get. *Arthritis and Rheumatism*, **37**, pp. 15-22.

Cherkin, D.C., Deyo, R.A., Wheeler, K. and Ciol, M.A., 1995, Physician views about treating low back pain. The results of a national survey. *Spine*, **20**, pp. 1-10.

Cherkin, D.C., Deyo, R.A., Street, J.H. and Barlow, W., 1996, Predicting poor outcomes for back pain seen in primary care using patients' own criteria. *Spine*, **21**, pp. 2900-2907.

Chew, C. and May, C., 1997, The benefits of back pain. *Family Practice*, **14**, pp. 461-465.

Chew-Graham, C. and May, C., 1999, Chronic low back pain in general practice: the challenge of the consultation. *Family Practice*, **16**, pp. 46-49.

Croft, P.R., Macfarlane, G.J., Papageorgiou, A.C., Thomas, E. and Silman, A.J., 1998, Outcome of low back pain in general practice: a prospective study. *British Medical Journal,* **316**, pp. 1356-1359.

Cypress, B., 1983, Characteristics of physician visits for back pain symptoms: a national perspective. *American Journal of Public Health*, **73**, pp. 389-395.

Deyo, R., 1996, Acute low back pain: a new paradigm for management. *British Medical Journal*, **313**, pp. 1343-1344.

Deyo, R., 1996, Drug therapy for back pain. Which drugs help which patients? *Spine*, **21**, pp. 2840-2849.

Deyo, R. and Diehl, A., 1986, Patient satisfaction with medical care for low-back pain. *Spine*, **11**, pp. 28-30.

Deyo, R., Diehl, A. and Rosenthal, M., 1987, Reducing roentgenography use. Can patient expectations be altered? *Archives of Internal Medicine*, **147**, pp. 141-145.

Dixon, A., 1990, The evolution of clinical policies. *Medical Care*, **28**, pp. 201-220.

Elam, K., Cherkin, D. and Deyo, R., 1995, How emergency physicians approach low back pain: choosing costly options. *The Journal of Emergency Medicine*, **13**, pp. 143-150.

Frank, J., Sinclair, S, Hogg-Johnson, S. Shannon, H., Bombardier, C, Beaton, D. and Cole, D., 1998, Preventing disability from work-related low-back pain. New evidence gives new hope – hope we can just get all the players onside. *Canadian Medical Association Journal,* **158**, pp. 1625-1631.

Freeborn, D., Shye, D., Mullooly, J.P., Eraker, S. and Romeo, J., 1997, Primary care physicians' use of lumbar spine imaging tests. Effects of guidelines and practice pattern feedback. *Journal of General Internal Medicine*, **12**, pp. 619-625.

Frost, H, Moffett, J.A.K., Moser, J.S. and Fairbank, J.C.T., 1995, Randomised controlled trial for evaluation of fitness programme for patients with chronic low back pain. *British Medical Journal*, **310**, pp. 151-154.

Grilli, R., Ramsay, C. and Minozzi, S., 2001, Mass media interventions: effects on health services utilisation. *The Cochrane Database of Systematic Reviews*, **1**.

Hart, L., Deyo, R. and Cherkin, D., 1995, Physician office visits for low back pain. Frequency, clinical evaluation, and treatment patterns from a U.S. national survey. *Spine*, **20**, pp. 11-19.

Hendler, N., 1984, Depression caused by chronic pain. *The Journal of Clinical Psychiatry*, **45**, pp. 30-36.

Hill, D., White, V., Marks, R. and Boland, R., 1993, Changes in sun-related attitudes and behaviours, and reduced sunburn prevalence in a population at high risk of melanoma. *European Journal of Cancer Prevention*, **2**, pp. 447-456.

Indahl, A., Velund, L. and Reikeraas, O., 1995, Good prognosis for low back pain when left untampered. A randomized clinical trial. *Spine*, **20**, pp. 473-477.

Industry Commission Report, 1995, Work safety and health: an inquiry into occupational health and safety. Commonwealth of Australia.

Jensen, M.C, Brant-Zawadzki, M.N., Obuchowski, N., Modic, M.T., Malkasian, D. and Ross, J.S., 1994, Magnetic resonance imaging of the lumbar spine in people without back pain. *New England Journal of Medicine*, **331**, pp. 69-73.

Klenerman, L., Slade, P.D., Stanley, M., Pennie, B., Reilly, J.P., Atchison, L.E., Troup, J.D.G. and Rose, M.J., 1995, The prediction of chronicity in patients with an acute attack of low back pain in a general practice setting. *Spine*, **20**, pp. 478-484.

Koes, B.W., Assendelft, W.J.J., van der Heijden, G.J.M.G. and Bouter, L.M., 1996, Spinal manipulation for low back pain. An updated systematic review of randomized clinical trials. *Spine*, **21**, pp. 2860-2871.

Kravitz, R., Callahan, E.J., Paterniti, D., Antonius, D., Dunham, M. and Lewis, C.E., 1997, Prevalence and sources of patients' unmet expectations for care. *Annals of Internal Medicine*, **125**, pp. 730-737.

Lethem, J., Slade, P.D., Troup, J.D. and Bentley, G., 1983, Outline of a fear avoidance model of exaggerated pain perception. Part 1. *Behaviour Research and Therapy*, **21**, pp. 401-408.

Linton, S.J., 2000, A review of psychological risk factors in back and neck pain. *Spine*, **25**, pp. 1148-1156.

Little, P., Smith, L., Cantrell, T., Chapman, J., Langridge, J. and Pickering, R., 1996, General practitioners' management of acute back pain: a survey of reported practice compared with clinical guidelines. *British Medical Journal*, **312**, pp. 485-488.

Loeser, J. and Sullivan, M., 1995, Disability in the chronic low back pain patient may be iatrogenic. *Pain Forum*, **4**, pp. 114-121.

Loisel, P., Abenhaim, L., Durand, P., Esdaile, J.M., Suissa, S., Gosselin, L., Simard, R., Turcotte, J. and Lemaire, J., 1997, A population-based, randomized clinical trial on back pain management. *Spine*, **22**, pp. 2911-2918.

Maetzel, A., Johnson, S.H., Woodbury, M. and Bombardier, C., 2000, Use of grade membership analysis to profile the practice styles of individual physicians in the management of acute low back pain. *Journal of Clinical Epidemiology.* **53**, pp. 195-205.

Malmivaara, A., Häkkinen, U., Aro, T., Heinrichs, ML., Koskenniemi, L., Kuosma, E., Lappi, S., Paloheimo, R., Servo, C., Vaaranen, V. and Hernberg, S., 1995, The treatment of acute low back pain – bed rest, exercises, or ordinary activity? *New England Journal of Medicine*, **332**, pp. 351-355.

Marks, R., 1999, Two decades of the public health approach to skin cancer control in Australia: Why, how, and where are we now? *Australasian Journal of Dermatology*, **40**, pp. 1-5.

Mathers, C. and Penn, R., 1999, Health system costs of injury, poisoning and musculo-skeletal disorders in Australia 1993-94. *Australian Institute of Health and Welfare: Canberra.*

Moffett, J.K., Torgerson, D., Bell-Syer, S., Jackson, D., Llewlyn-Phillips, H., Farrin, A. and Barber, J., 1999, Randomised controlled trial of exercise for low back pain: clinical outcomes, costs, and preferences. Measuring the functional status of patients with low back pain. Assessment of the quality of four disease-specific questionnaires. Low back pain: which is the best way forward? *British Medical Journal*, **319**, pp. 279-283.

Nachemson, A., Waddell, G. and Norlund, 2000, A. *Epidemiology of neck and back pain*, in *neck and back pain. The scientific evidence of causes, diagnosis, and treatment*, edited by. Nachemson, A. and Jonsson, E., (Philadelphia: Lipppincott Williams & Wilkins), pp. 165-188.

Owen, J.P., Rutt, G., Keir, M.J., Spencer, H., Richardson, D., Richardson, A. and Barclay, C., 1990, Survey of general practitioners' opinions on the role of radiology in patients with low back pain. *British Journal of General Practice*, **40**, pp. 98-101.

Owen, N., Bauman, A., Booth, M., Oldenburg, B. and Magnus, P. 1995, Serial mass-media campaigns to promote physical activity: reinforcing or redundant? *American Journal of Public Health*, **85**, pp. 244-248.

Philips, H. and Jahanshahi, M., 1986, The components of pain behaviour report. *Behaviour Research and Therapy*, **24**, pp. 117-125.

Phillips, D., Kanter, E.S., Bednarczyk, B. and Tastad, P.L., 1991, Importance of the lay press in the transmission of medical knowledge to the scientific community. *New England Journal of Medicine*, **325**, pp. 1180-1183.

Pierce, J., Macaskill, P. and Hill, D., 1990, Long-term effectiveness of mass media led anti-smoking campaigns in Australia. *American Journal of Public Health*, **80**, pp. 565-569.

Prochaska, J. and DiClemente, C., 1982, Transtheoretical therapy toward a more integrative model of change. *Psychotherapy: Theory, Research and Practice*, **19**, pp. 276-287.

Redman, S., Spencer, E. and Sanson-Fisher, R., 1990, The role of mass media in changing health-related behaviour: a critical appraisal of two methods. *Health Promotion International,* **5**, pp. 85-101.

Roland, M. and Dixon, M., 1989, Randomised controlled trial of an educational booklet for patients presenting with back pain in general practice. *Journal of the Royal College of General Practitioners*, **39**, pp. 244-246.

Roland, M. and van Tulder, M., 1998, Should radiologists change the way they report plain radiography of the spine? *The Lancet*, **352**, pp. 229-230.

Roland, M., *et al.*, 1996, *The Back Book*. (United Kingdom: The Stationary Office).

Rose, G., 1992, *The strategy of preventive medicine*. (Oxford: Oxford University Press).

Rossignol, M., Abenhaim, L., Séguin, P., Neveu, A., Collet, JP., Ducruet, T. and Shapiro, S., 2000, Coordination of primary health care for back pain. A randomized controlled trial. *Spine*, **25**, pp. 251-259.

Schiottz-Christensen, B., Neilsen, G.L., Hansen, V.K., Schodt, T., Sorensen, H.T. and Olesen, F., 1999, Long-term prognosis of acute low back pain in patients seen in general practice: a 1-year prospective follow-up study. *Family Practice*, **16**, pp. 223-232.

Schroth, W.S, Schectman, J.M., Elinsky, E.G. and Panagides, J.C., 1992, Utilization of medical services for the treatment of acute low back pain: conformance with clinical guidelines. *Journal of General Internal Medicine*, **7**, pp. 486-491.

Skelton, A., Murphy, E.A., Murphy, R.S. and O'Dowd, T.C., 1996, Patients' views of low back pain and its management in general practice. *British Journal of General Practice*, **46**, pp. 153-156.

Sly, D.F, Hopkins, R.S., Trapido, E. and Ray, S., 2001, Influence of a counteradvertising media campaign on initiation of smoking: The Florida "truth" Campaign. *American Journal of Public Health*, **91**, pp. 233-238.

Symonds, T.L., Burton, A.K., Tillotson, K.M. and Main, C.J., 1995, Absence resulting from low back trouble can be reduced by psychosocial intervention at the work place. *Spine*, **20**, pp. 2738-2744.

Symonds, T., Burton, A.K., Tillotson, K.M. and Main, C.J., 1996, Do attitudes and beliefs influence work loss due to low back trouble? *Occupational Medicine*, **46**, pp. 25-32.

van den Hoogen, H., Koes, B.W., van Fijk, Th.M. and Bouter, L.M.,1998, On the course of low back pain in general practice: a one year follow up study. *Annals of the Rheumatic Diseases*, **57**, pp. 13-19.

van der Weide, Verbeek, W.J. and van Tulder, M., 1997, Vocational outcome of intervention for low-back pain. *Scandinavian Journal of Work, Environment and Health*, **23**, pp. 165-178.

van Tulder, M., Koes, B. and Bouter, L., 1997a, Conservative treatment of acute and chronic nonspecific low back pain. A systematic review of randomized controlled trials of the most common interventions. *Spine*, **22**, pp. 2128-2156.

van Tulder, M.W., Assendelft, J.J., Koes, B. and Bouter, L., 1997b, Spinal radiographic findings and nonspecific low back pain. A systematic review of observational studies. *Spine*, **22**, pp. 427-434.

Victorian WorkCover Authority, 1996, *Guidelines for the management of employees with compensable low back pain*.

Victorian WorkCover Authority, 2001, In *Annual report Victorian WorkCover Authority 1999/2000*. Melbourne (Vic).

von Korff, M., 1994, Studying the natural history of back pain. *Spine*, **19**, pp. 2041-2045.

von Korff, M., Barlow, W., Cherkin, D. and Deyo, R., 1994, Effects of practice style in managing back pain. *Annals of Internal Medicine*, **121**, pp. 187-195.

Waddell, G., 1996, Low back pain: a twentieth century health care enigma [Keynote address for primary care forum]. *Spine*, **21**, pp. 2820-2825.

Waddell, G., 1987, 1987 Volvo award in clinical sciences. A new clinical model for the treatment of low-back pain. *Spine*, **12**, pp. 632-644.

Waddell, G., Newton, M., Henderson, I., Somerville, D. and Main, C.J., 1993, A fear-avoidance beliefs questionnaire (FABQ) and the role of fear-avoidance beliefs in chronic low back pain and disability. *Pain*, **52**, pp. 157-168.

Waddell, G., Feder, G. and Lewis, M., 1997, Systematic reviews of bed rest and advice to stay active for acute low back pain. *British Journal of General Practice*, **47**, pp. 647-652.

Wensig, M., van der Weijden, T. and Grol, R., 1998, Implementing guidelines and innovations in general practice: which interventions are effective? *British Journal of General Practice*, **48**, pp. 991-997.

Wiesel, S.W., Tsourmas N., Feffer, H.L., Citrin, C.M. and Patronas, N., 1984, A study of computer-assisted tomography. The incidence of positive CAT scans in an asymptomatic group of patients. *Spine*, **9**, pp. 549-551.

Zarkowska, E. and Philips, H., 1986, Recent onset vs. persistent pain: evidence for a distinction. *Pain*, **25**, pp. 365-72.

CHAPTER 7

Stakeholder Engagement in the Control of Repetitive Strain Injury

Michael Polanyi and Donald Cole

7.1 INTRODUCTION

A "new epidemic" of diverse work-related injuries and illnesses emerged during the latter part of the twentieth century (Rantanen *et al.*, 1994). They include musculoskeletal injuries of the neck, arms and back, work related musculoskeletal disorders (WMSD), fibromyalgia and chemical sensitivities among others. In earlier work (Polanyi, 2001), we have characterized WMSD as "wicked problems" (Mason and Mitroff, 1981) based on several attibutes: they are difficult to diagnose objectively and thus are *uncertain* and *value-laden*; they are *systemic* since they are caused by a range of individual, work and societal factors; and they are *conflict-ridden* since they impact on numerous and sometimes competing groups (workers, employers, government, taxpayers).

Stakeholder responses to such injuries and illnesses have been conflicting, particularly in North America. Many employers have questioned the legitimacy of these injuries and illnesses, denying the causal contribution of work activities and conditions. Some employers have been more pro-active, primarily implementing work-station level changes to address WMSD, with mixed success. Unions have increasingly considered soft tissue injuries to be fundamentally caused by the intensification of work demands and the stripping away of health and safety protections due to increases in employer power (and declines in unionization in the USA) during the 1980s and 1990s. They have argued for the implementation of new "ergonomic standards", an uphill struggle, only partially successful in an anti-regulatory environment. North American governments have generally been unable to facilitate a shared vision and agenda to respond to musculoskeletal injuries. Researchers have sought to reconcile workplace party differences by providing neutral and conclusive evidence about diagnosis, causality and prevention.

Mason and Mitroff (1981) have argued that broad participation on "wicked" issues is required since: the "raw material" for problem solving is dispersed across a range of stakeholders; some form of collective risk sharing is needed to deal with the consequences of wicked problems; and effective policy is made *with* not *for* or *at* stakeholders. For WMSD, the time seems ripe for multi-stakeholder action for several reasons. First, health and safety has been a fruitful area of union-management cooperation, thanks to the legislated right to participate that exists in many jurisdictions. Second, the importance of full and broad participation in successful change processes is being increasingly recognized in a range of domains including community health (Epp, 1986), planning (Forester, 1999), organizational change (Neuman *et al.*, 1989), research dissemination (National Center for the Dissemination of Disability Research, 1996), and, more specifically, in the domain of workplace health and ergonomics interventions (Moore and Garg, 1997). Third, workplace parties have increasingly become involved as collaborators with researchers, helping to define research questions, carry out research activities, and interpret and apply research findings, as exemplified in other contributions to this book.

The two case studies described in this chapter build upon a long tradition of researcher-workplace party "action research", a process of collective inquiry and problem solving which aims to generate new knowledge and solve real-life problems. Action researchers assume that the separate knowledge of outside professional researchers and inside practitioners is incomplete and open to different interpretations (Evered and Louis, 1981). They argue that "insiders" and "outsiders" need to research, learn and interact together to reach a deeper understanding of complex social phenomena. Moreover, since action and change are shaped as much by values and resources as by knowledge, action researchers believe that change processes require the participation of a range of actors, both scientific and action oriented. Action researchers tend to focus on building common understanding to incrementally address shared social problems. Various forms of action research have been applied to worker health concerns in different countries: job redesign workshops in Norway (Gustavsen, 1996), the "LOM" program in Sweden (Gustavsen, 1996), participatory research in Quebec (Mergler, 1987), and action research (Pasmore and Friedlander, 1982) and "participatory action research" (Israel *et al.*, 1992) in manufacturing plants in the United States.

We believe action research offers several potential advantages over conventional research for the stimulation of research-informed change on complex workplace health issues. It acknowledges there are multiple dimensions of "truth" and has the potential to enhance participation in research and change processes. In this chapter, we discuss lessons from two action/research-based interventions aimed at reducing the burden of upper extremity WMSD, commonly known as repetitive strain injuries (RSI). The cases – one within a single workplace, and one at the regional level – illustrate both the opportunities and the limitations of researcher-workplace processes to improve worker health. A number of tensions and trade-offs are revealed and explored. We end by proposing a framework to enhance multi-stakeholder collaboration on workplace health concerns.

7.2 A MULTI-STAKEHOLDER "FUTURE SEARCH" CONFERENCE ON RSI

7.2.1 Origins and Methods

The first case study is of a regional Future Search conference, implemented to build "common ground" and stimulate multi-stakeholder action on RSI/WMSD.

Future Search is a participatory planning approach aimed at building common visions and action on difficult organizational and community issues (Weisbord and Janoff, 1995). It differs from conventional planning approaches because it involves the "whole system" of stakeholders, it focuses on the context of the issue, it avoids problem solving, and it relies on a large degree of self-management by participants. The Future Search arose from a perceived need for greater stakeholder interaction in order to build mutual understanding and a more common vision for the prevention and treatment of RSI. Initial response from key players was positive, and by the fall of 1997 one of us (Polanyi) had pulled together a multi-stakeholder design team that agreed to undertake a Future Search conference on RSI with technical support by two experienced consultants and sponsorship by the Institute for Work & Health (IWH). Design team members liked the interactive nature of the Future Search process, its potential to break down barriers among stakeholders and the opportunity to bridge gaps in stakeholder positions. Further, they thought it could increase participation on RSI and inform practice by current research.

The conference was held over three days in May 1998, involving some 60 stakeholder representatives from southern and eastern Ontario (there were two international participants). At the conference, participants followed closely the highly structured Future Search process which involves four main stages: developing a shared history, identifying key trends, developing a future vision, and then identifying common ground and action plans. Documentation and evaluation of the process and outcomes of the conference were carried out primarily through a detailed qualitative analysis of conference materials, planning meetings and post-conference participant interviews and questionnaires (details in Polanyi, 2001).

7.2.2 Process and Outcomes

Securing an inclusive, equal and non-coercive process

The RSI Future Search achieved significant attendance from all key stakeholder groups. Several reasons may have lead to such success. First, it had a broad-based, informed and well-connected design team. Team members had a good knowledge of the various dimensions of RSI and of the key players within each stakeholder group. Second, a systematic and detailed process of determining stakeholder groups and potential participants within each stakeholder group was followed. The facilitators emphasized the importance of having diversity within each stakeholder group, and helped participants develop an explicit criteria for selection (e.g. age, location, unionization etc.). Third, design team members "worked their connections", spent a lot of time advocating for the participation of key players, and did not succumb to our initial sense that people would not come to a meeting for three days.

Despite all these efforts, certain stakeholder groups, and participants with certain characteristics were under-represented. Union representatives, representatives from related government departments (finance, health), and, above all, employer representatives with responsibilities beyond health and safety, were lacking. In a sense, this made discussion easier, as union-management conflict remained largely below the surface. But this also meant discussions were limited in scope. Second, there was some lack of diversity in cultural background, age and socio-economic status. There were clearly many white, middle-aged, middle-class, urban professionals in the room, partly a reflection of who occupies professional positions in occupational health and safety but also of the demographics of design team members. Moreover, the decision to combine injured workers *and* their professional advocates in a stakeholder group limited the number of injured workers present and may have limited socio-economic diversity.

Beyond balancing of attendance, ensuring equal engagement of all participants can be problematic. Future Search proponents claim that the process allows participants to work as peers, and much emphasis is placed on setting up a non-judgmental, accepting climate for discussion. Name tags were devoid of titles and positions, there were no invited "experts" sitting on a raised stage, and facilitators emphasize that all participants have important experience and knowledge to contribute. Just being in the same room with other stakeholders for an extended period, hearing the experiences and perspectives of others, and working in small cross-stakeholder groups certainly encouraged some participants to revise their stereotypes.

There was some breadth of involvement: over 80% of participants contributed at least once in the large group sessions. Yet imbalances still persisted. In plenary sessions, each member of the four most vocal stakeholder groups (researchers and educators, injured worker representatives, ergonomists and health professionals) spoke 40 times on average, whereas each member of the remaining five stakeholder groups spoke only 11 times. The four most vocal individuals averaged 19 contributions each, the next four most vocal participants averaged nine contributions.

Participants identified some barriers to participation: intimidation in face of the "eminent scientists" present; poorly functioning small groups where the extroverts dominated; and a lack of willingness to listen. Speaking in front of a group of 60 peers can be more of a challenge than is recognized. Some participants seemed to thrive in the large group sessions, emerging as leaders with the capacity to speak articulately in a large group and make timely contributions. On the other hand, there appeared to be some anger and withdrawal on the part of participating injured workers, who did not feel their views were being understood. Indeed, participating injured workers and their representatives highlighted the fact that injured workers have the least power to bring about change on RSI, while at the same time experiencing most directly the harsh physical, financial and emotional impacts of these injuries.

Reaching Common Ground and Agreeing On Action

At the conference, participants established a degree of shared understanding of RSI/WMSD (see Polanyi, 2001). It was agreed that RSI is complex and not easily defined, and that both work- and non-work factors contribute to RSI. It was also agreed that more research on RSI, better sharing of information about the impacts of RSI, and sharing of "best practices" for prevention and treatment of RSI were needed. A common sense of the importance of dialogue and cooperation among different stakeholders also emerged.

On the other hand, differences in views persisted about how RSI and workplace health are being affected by broader political, economic and technological changes; about what workers and employers should expect and be responsible for in today's world; and about whether the best approach to dealing with RSI is voluntary or regulatory. Divergent responses to these questions revolved around opposing views of the relevance and impacts of economic globalization and greater international competition. Some felt globalization was stimulating a new valuing of human resources; others felt it was leading to deregulation, increased work pace and intensity, and reduction of employment and health and safety standards. However, Future Search participants did not follow up on these identified tensions between firm competitiveness and worker health. Rather, the Future Search method seems to have encouraged participants to *assume* that workplace health and safety and RSI are phenomena marked more by consensus than by conflict. The focus on commonality helped prevent the process from getting bogged down in debate, especially as the conference moved toward action, but it also left uncertainty as to what was actually agreed upon.

A number of action themes were identified by volunteer "champions" but the process of arriving at them was problematic. The boundary between the long and undeveloped lists of common ground themes and unresolved differences remained unclear, despite clarification efforts. It also remained unresolved how much action plans had to be rooted in the common ground identified, and whether action champions were free to go wherever their passions and energy took them. When one action group proposed a plan to advocate for ergonomic regulations, the whole group split in two (as has occurred in the US debates – OSHA 2000). Some felt that ergonomic regulations were an "unresolved difference". Others felt that *sub-groups* had the right to do what they wanted, as long as they did not do it in the name of the *whole* group.

Implementing Action Plans

The conference stimulated four kinds of action: presentations and exchange of information; networking or mutual support; new projects and initiatives; and broader organizational changes (see Polanyi 2001 for details). *Information/ educational activities*, often involving a variety of stakeholders, included over ten presentations and articles for a range of audiences: ergonomists and health and safety practitioners, clinicians, psychologists, labor representatives, organizational consultants, and compensation officials. In addition a computer-based listserve on RSI in Ontario was set up. While *networking* is often stimulated at conferences, both the level and scope of cooperation were notable. Participants made contacts with people they felt they would not have otherwise met. There was cooperation of various kinds: an employer received support from a government office for injured workers, a consultant sought participation of injured workers in teaching a course, a doctor wrote a letter on behalf of an injured worker, and a compensation board employee and a researcher made a joint presentation to the compensation board.

Some *new initiatives* were launched as a result of connections made and ideas discussed at the conference. For example, a project to involve injured workers as researchers on compensation and return to work issues was submitted to and subsequently funded for two years by the Ontario compensation authority, the Workplace Safety & Insurance Board (WSIB). Another project to allow a multi-stakeholder group to visit workplaces successfully dealing with RSI risk factors was developed. The conference also fed into various *organizational decisions*. For example, following the conference, the WSIB proposed a Future Search style process on occupational disease, in general. The RSI Future Search also informed decisions being made in several areas: the development of best practices criteria at the WSIB; the content of a major research proposal on RSI; and a shift toward more stakeholder interaction by IWH.

Key barriers to the fulfillment of actions plans cited by participants included lack of available time. Indeed, some action-group leaders and group members faced heavy and competing work demands. The more "active" groups all had "champions" who were highly motivated and involved, partly because the goals of their action groups dove-tailed with their own personal, professional or work-related goals. Other forms of staff support were therefore procured, such as funding for a part-time coordinator. Further, too broad action group goals were problematic. For example, the plans of the education group and health and productivity groups aimed to scope out current activities in their respective areas. In contrast, goals of the more active groups tended to be specific, limited and directly achievable. Nevertheless, three years after the conference, a number of participants continued to collaborate on different projects and refer back to their experiences in the RSI Future Search process. One of the locations of such collaboration was in the workplace-based case that follows.

7.3 COLLABORATIVE, WORKPLACE-BASED RESEARCH ON RSI

7.3.1 Development

During the early 1990s, "outbreaks" of RSI had occurred in different departments of a large, metropolitan newspaper, *The Toronto Star*. Despite various initiatives to deal with RSI, lost-time claims were increasing in the mid-1990s. Members of the Southern Ontario Newspaper Guild (SONG), the main union at 'The Star', were increasingly upset about the impact of RSI on their members, deeming it one of the consequences of the introduction of technology without adequate safeguards (Schenk and Anderson, 1995). In early 1995, they approached IWH to become involved in research on RSI.

7.3.2 Participation

During negotiations of a new collective agreement, union and management agreed to partially fund RSI research. An "RSI Watch" committee was constituted, following a tripartite model, with equal numbers of management and union representatives chosen by their respective constituencies and at least one researcher present. Researcher numbers varied depending on the content of the meeting and the need for additional input. Being sensitive to the desire for some equality of voice between the parities, we attempted to restrict attendance to numbers roughly equal to those participating from union and management (between two and four, depending on the meeting). This steering committee became the key form of joint workplace party-researcher participation. It was transformed into an RSI Committee in the subsequent round of collective bargaining three years later and continues at the time of writing.

7.3.3 Formulation of Objectives

Based on discussions with union and management and the first RSI Watch meeting, the researchers developed a set of objectives for the action research. These were subsequently modified in the RSI Watch Committee to four objectives:

1) to identify and examine the interplay between individual, biomechanical and work organization factors related to the cause and course of RSIs at *The Toronto Star*
2) to determine and seek evidence for the effectiveness of current preventive and rehabilitative interventions practiced at/by *The Toronto Star*
3) to recommend organizational, biomechanical and rehabilitative interventions to reduce the impact of RSIs, based on the results of objectives 1 and 2
4) to evaluate the effectiveness of such interventions.

7.3.4 Research Implementation

The workplace parties' primary interest in identifying the full breadth of the problem led to a cross-sectional survey of all office employees at the newspaper in 1996 (Polanyi *et al.*, 1997). Questions were discussed at RSI Watch Committee meetings and newspaper staff assisted with graphics for the questionnaire. Union and management members arranged piloting with members/colleagues at other employers. Workplace parties arranged questionnaire distribution and confidential collection through workplace health and safety representatives and United Way captains. Data tabulation and preliminary analyses carried out by the research team were shared with the workplace parties in RSI Watch meetings. Subsequent interest of both workplace parties and researchers in more detailed information on some of the themes arising from the survey led to a series of sub-studies in the first half of 1997, both qualitative (e.g. in-depth interviews and focus groups) and quantitative (e.g. physical assessments, direct observations, questionnaire on treatment). Workplace parties played important roles in interpreting findings, guiding further analyses and formulating ways of presenting findings to others in the workplace. Articles were prepared for "Star Beat", the in-house newspaper.

7.3.5 Deciding on Action

Although workplace parties took some actions based on research findings discussed at RSI Watch Committee meetings, the major thrust occurred during a series of RSI Watch Committee workshops in the fall of 1997. Different working groups brought together the research results at the Star, the experience of workplace parties and researchers in dealing with RSI/WMSD, and knowledge of research and programs from elsewhere, to develop a set of recommendations. These were embedded in both presentations and a research team report to executives of the union and the company. The union put together a proposal to management, incorporating many of the recommendations. Key management leaders persuaded executive management to support the recommendations and include them in a letter of understanding as part of contract negotiations.

7.3.6 Implementation of recommendations on action and evaluation

The RSI Watch Committee metamorphosed into the RSI Committee, with some modifications in membership, in the spring of 1998. The new RSI Committee has formal status, with minutes kept by the workplace parties and a full action agenda. Research-related issues are discussed during the second half of the meetings, in keeping with a modified researcher role as best practice and evaluation consultants. The RSI Committee developed a wide ranging Ergonomic Policy and implemented a "Stop RSI" program, including an awareness campaign to encourage reporting, mandatory training, workstation assessments, onsite physiotherapy and problem solving responses when concerns about RSI/WMSD arise in different departments (see Wells *et al.*, 2001, for more detail).

7.3.7 Learning about dealing with RSI/WMSD in workplaces

Through participation in meetings, interviews with members of the RSI Watch and RSI Committees, review of Joint Health and Safety Committee meeting minutes and interactions with employees during different projects over the course of our collaborative research, we have been able to better understand a number of issues to tackle in order to reduce the burden of RSI/WMSD. Here we review them, roughly ranked according to the ease with which the mid-level workplace parties, with whom we deal as researchers, appear able to reach consensus and bring about change.

Perhaps easiest is *awareness*. At one time, many employees had only heard about RSI through hearing or reading stories on high-profile cases among Editorial staff. Since our involvement and the hiring of a health and safety specialist with training in ergonomics, stories on RSI have appeared in "Star Beat", posters have been produced about the symptoms of RSI, the full range of employees including managers have undergone mandatory training, and orientation materials, including a video, have been developed for new employees. Many employees now appear to know what RSI is and to be aware that work can contribute to the development of RSI or aggravate an existing RSI. As one RSI Committee member commented: "Now everybody [is] admitting that it's a problem".

Deepening *understanding* of the causes and impacts of RSI and changing *attitudes* has been harder. Many who face risks for RSI (e.g. inadequate workstations), yet have not developed RSI, have found it hard to accept that they might be at risk. This sense of invulnerability often comes with a skepticism directed at RSI sufferers. Despite evidence from the research that those with greater frequency, duration and intensity of symptoms are more likely to experience limitations in carrying out tasks at work and activities of daily living (Beaton *et al.*, 2000), some find it hard to understand why some employees with RSI should require extensive accommodation of duties and/or equipment in order to remain working. Partly this may be related to the wide range of symptoms and impacts that people experience, best visualized as an iceberg (see Figure 7.1). At the base were the broadest group with some pain/discomfort, only some of whom found that it was aggravated by work, fewer of whom sought healthcare and even fewer of whom had lost work days associated with RSI.

Perhaps because of such variability, individuals differed considerably in their *willingness to report* RSI to the workplace. With the provision of onsite physiotherapy at the largest office complex of The Star, many more employees came forward for healthcare. The treating physiotherapist was struck by the extent to which those she treated initially had had RSI for a long time. Most clients are now informing their supervisor as well as the health center staff, in keeping with one of the aims of the "Stop RSI" program, namely increased reporting by those with symptoms aggravated by work.

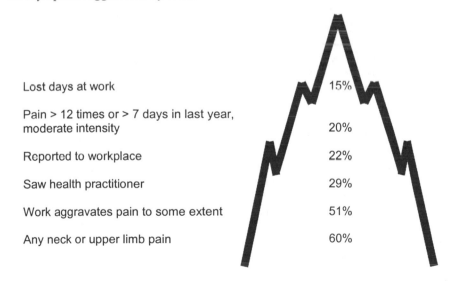

Lost days at work	15%
Pain > 12 times or > 7 days in last year, moderate intensity	20%
Reported to workplace	22%
Saw health practitioner	29%
Work aggravates pain to some extent	51%
Any neck or upper limb pain	60%

© Institute for Work & Health, 1999

Figure 7.1 The "Iceberg" of WMSD.
One-year prevalence of WMSD at a large newspaper company in 1996.

The adoption of personal strategies to prevent and to manage RSI also varies considerably. Placement of workstation components and postures adopted during working can be affected as much by past habits as current workstation configurations and tasks. Strong personal commitments to complete work, organizational requirements to meet deadlines and incentive structures to increase productivity, all lead to employees focusing so hard on work that they may not take breaks. RSI Committee members have expressed frustration at the difficulties in changing such habits or practices. An analogy has been drawn to the debate around who is responsible for wearing personal protective equipment, e.g. a hard hat on a construction site. Labour representatives have tended to argue that it is the responsibility of supervisors and managers to make sure that such practices are adhered to. Management representatives have more often argued that employees are adults who make their own decisions. Both agree on the importance of reminding both employees and managers on a regular basis, with personal observation and guidance whenever possible.

Social interactions of different kinds are immensely important in both preventing and dealing with RSI. Our cross-sectional survey documented that those who felt more social support at work were less likely to experience a more severe RSI (Polanyi *et al.*, 1997). How one's manager reacts can have a major impact on an employee's ability to get help and to adapt their workstation and work tasks to remain productive. In some departments, such as Editorial, a culture of sharing has developed which promotes people telling and supporting others, particularly co-workers and the management staff person with part-time responsibilities for RSI.

Organizational activities have also taken place. Trained employees have conducted ergonomic assessments of employees at their workstations though at a slower pace than most RSI Committee members would like, due to the heavy time pressures that assessors face while they carry on their regular jobs. The development of policies around equipment purchasing, and input on the choices during major moves and renovations, have been an important activity of RSI Committee members, particularly the management health and safety staff person and the management representative in charge of the move, who was expressly appointed to the RSI Committee for this purpose. Unfortunately, cost and space considerations were also important decision factors resulting in less than optimal choices being made from an RSI perspective. Input has also been provided to the Future Systems group around design of new software to be used in several departments over the coming years. Unfortunately, although high demands at work or workload are understood to be risk factors for RSI, staffing levels which impact on workload tend to be regarded by both parties as labor relations issues. To deal with concerns that inadequate staffing on the copy-editing desk was increasing the risk of RSI for copy editors, the labor RSI Committee representative had to push for a meeting with the responsible Editorial manager in order to achieve adoption of measures for adequate coverage of staff leaves to maintain manageable workloads. In these ways, RSI Committee members must interact with other organizational decision-makers to bring about broader changes.

7.4 COMMON BENEFITS AND CHALLENGES OF MULTI-STAKEHOLDER PROCESSES

A great challenge in multi-stakeholder processes is the multiplicity of inter-related (and sometimes conflicting) goals. Multi-stakeholder processes seek, to varying degrees, to secure active involvement, encourage learning, forge common ground and stimulate action. Some of these aims are compatible: broad participation and the incorporation of diverse positions *can* help lead to a fuller understanding of complex issues and a greater commitment to action. Other aims are more in conflict: creating conditions for a learning process can sometimes conflict with goals of inclusiveness and action. Although our two case studies involved quite different processes, they were carried out on the same problem (RSI/WMSD) and reveal several common benefits and challenges involved in multi-stakeholder processes.

7.4.1 Generating Participation and Commitment

In both cases, broad participation and involvement generated a high level of commitment among participants. In the RSI Future Search process, the fact that a range of participants are still working to build multi-stakeholder action three and a half years after the start of the process is a testimony to the level of commitment generated. In the workplace case, high levels of commitment to understanding and dealing with RSI have been generated among important stakeholders: employees, union, management and the WSIB. Through the series of studies and the Stop RSI program activities, a large number of employees have been involved in some way, with direct exposure to information on RSI, a key to changing attitudes and practices. One of the clearest examples was how one participant who was wired up for measurement of physical exposure and had video taken of her activities spoke about how she became a walking advertisement for the "RSI thing". As well, some of the most consistent comments during key informant interviews were on "the enthusiasm shown by top management and human resources" and "how quickly people were excited about" the Stop RSI program.

7.4.2 Balancing Shared Learning and Scientific Rigor

One tension in multi-stakeholder processes exists at the interface between knowledge and evidence, on one hand, and participation and voice, on the other. How can one *inform* multi-stakeholder processes with current research evidence, without subtly dismissing other ways of knowing about complex issues, such as first-hand experiences? How does one involve experts without restricting the inclusiveness of the process?

Future Search designers feel that presentations by experts create hierarchy and hinder collective learning, and that research knowledge can be adequately brought into the process by the presence of researchers as *participants*, rather than as presenters, so no formal presentations were included. Although conference participants felt good about the level of interaction and collaboration, some lamented the lack of an opportunity to hear from the researchers in the room about the current state of knowledge. They also regretted not hearing about current initiatives on RSI, in order to avoid duplication in their action plans.

In the workplace action research case, the research team aimed at promoting a learning environment. At times, we acted as advocates for the importance of particular perspectives based either on other research, such as the inclusion of work organization questions on the cross-sectional questionnaire, or on our own findings, such as the process by which people define "when they have a problem with RSI". The workplace parties brought the richness of their own experience and knowledge about RSI in their workplace to bear in ways that the research team had not envisaged, thus improving validity and rigor overall.

7.4.3 Promoting Consensus and Accepting Conflict

Workplace health issues exist at a complex interface of the interests of various stakeholder groups which are partly aligned and partly in conflict (see Brown, 1983, on organizational interfaces). On the one hand, employers and workers share an interest in the success of the firm, and the employer wants healthy employees who can contribute productively to that success. On the other hand, workers and employers are in conflict to the extent that employers want to get as much out of workers *at the least cost*, and to *avoid* having to invest in health and safety, ergonomic equipment, etc. Broader economic and political incentives and regulations can encourage employers to take the "high road" of investing in and supporting long-term employee productivity, as opposed to the "low road" of short-term cost-cutting. Discussions at the Future Search conference did not fully explore such tensions in context. The emphasis on creating an accepting and inclusive environment and focusing on common ground rather than difference in the Future Search method, seems to have made participants hesitant to challenge or question other participants. A core danger in multi-stakeholder processes is that by involving all stakeholder groups in the process, the implicit message can be that *all* the interests of *all* stakeholder groups are legitimate *and* that all these interests are, fundamentally, reconcilable (see Polanyi, 2002). It appears that clarification of the boundaries between commonality and difference, the factors affecting the location of those boundaries, and the impacts of difference on action is an ongoing requirement in multi-stakeholder processes.

7.4.4 Role of Researchers

In the workplace case, discussions during RSI Watch and RSI Committee meetings about what to include in the research, how to interpret results and how to act on them varied from consensual to conflictual. At times, we acted as mediators between divergent opinions, such as when staffing levels were put forth as risk factors by union members and management members felt uncomfortable as to whether the research spoke to the issue. During the protracted periods of discussion of the research findings, both qualitative and quantitative, RSI Watch and RSI Committee members often argued fiercely over interpretation. As researchers we sometimes had to take a back seat when hotly contested workplace issues polarized union and management representatives, such as when a team re-organization process was modified by management in one department. At other times, RSI Committee members surged forward with actions, such as training and workstation assessment activities, in ways that may not have taken full advantage of literatures relevant to their design. On broader organizational policies, such as incentive structures and workflow design, which our research indicates are important for WMSD/RSI, neither we nor the RSI Committee appear to have sufficient leverage to influence what are regarded as "management prerogatives".

7.4.5 Trade-offs between Learning/Reflection and Intervention/Action

In action research processes involving multiple stakeholders, decisions must be made as to resources and time devoted primarily to learning/reflection versus intervention/actions. Critical reflection and analysis of RSI seems to have been limited by various aspects of the structure of the Future Search process. While a wide variety of perspectives of the causes and consequences of RSI/WMSD arose early in the conference, the subsequent rush to envision ideal futures and to develop action plans within the tight time frame available meant that these perspectives were not subject to further reflection or analysis. A focus on "easier actions" – education, information sharing, multi-stakeholder dialogue – resulted. These kind of actions (voluntary, informative) rest on the assumption of a fundamental commonality of worker, employer and government interests. Yet multi-stakeholder processes that assume *prima facie* a consensus model of the issue at hand in order to achieve action plans are in danger of ignoring important determinants of workplace health problems and perpetuating the status quo.

The workplace research process has also taken much longer than virtually all the workplace parties anticipated. Research has resulted in heavy time demands on RSI Committee members and employees who participate in study components. Similarly our concerns for confidentiality, which is crucial in workplace settings, given the private way in which most of us understand our health, at times became a block to action. When we found that some employees' workstations were still not adjusted several months after a move to new workstations, we could only encourage the affected participants to approach staff to obtain earlier assessments and encourage the same staff to broadly initiate such assessments. Yet overall, both workplace parties have appreciated how having an ongoing action research program has provided evidence to back up their RSI concerns, has kept RSI on the broader workplace agenda, and has provided an impetus for policy and program initiatives they wanted to take.

7.5 TOWARD A FRAMEWORK FOR MULTI-STAKEHOLDER ACTION ON COMPLEX WORKPLACE HEALTH ISSUES

Promoting research-informed action on WMSD and other complex workplace health issues is a challenge. Yet complex, multi-causal workplace injuries and illnesses will not go away, so we need to tackle the challenge of stimulating research-informed action as Frank and Sullivan argue in the opening chapter. Ongoing negotiation around and flexibility in researcher roles seems to be a key requirement for researcher-stakeholder engagement that can achieve both learning and action. The above case studies suggest that research-practitioner collaboration is both desirable and possible.

During the 1960s and 1970s, unions and other activists struggled successfully to entrench the new rights to full knowledge and participation on workplace health and safety issues. Yet challenges remain in improving the *quality* of knowledge and participation, and in developing effective techniques to ensure that knowledge and participation translate into improved workplace conditions and worker health. In some ways, thinking about the nature of participation in workplace health and safety has fallen behind that in other fields like organizational development (Neuman *et al.*, 1989), community-based planning (Forester, 1999), and health promotion (Labonte, 1994). Those concerned with worker health and safety may also now learn from the environmental field, where a range of creative approaches to citizen participation have been developed, such as citizen advisory committees, planning cells and citizen juries (Renn *et al.*, 1995). Based on an assessment of these approaches, Thomas Webler (1995) has identified two core dimensions of "right discourse" in citizen participation: *competence* and *fairness*.

To Webler, *competence* requires that all participants are provided with "the procedural tools and knowledge needed to make the best possible decisions". This means all participants must have equal and adequate access to knowledge and interpretations and that the best procedures be used for resolving disputes about knowledge and interpretation. Preparatory work with the least powerful stakeholder groups may be required to build their knowledge base, capacity and depth of analysis *before* they enter multi-stakeholder processes.

Fairness requires "each participant should feel safe and comfortable about advancing any type of statement and participating in judging the validity of others' claims" (Webler, 1995). This means all potentially affected by the decision have an equal chance to be present or represented; all have the opportunity to put their concerns on the agenda and to propose or approve rules for discourse and facilitation; and all have an equal chance to put forth and criticize – and help resolve differences among – claims about language, facts, norms and expressions.

We believe that neither competence nor fairness *on its own* is sufficient. So, while we support the spirit of the recent swell of support for "evidence-based decision-making" as one means of ensuring competence of decision making, we feel that it does not pay sufficient attention to the fairness of the decision making process. On the other hand, we believe that some collective bargaining processes and political processes, rooted in a discourse of justice, could benefit from a stronger grounding in knowledge generated through research.

Striking a balance between competence and fairness is a challenge. The Future Search conference described here erred on the side of fairness: there was extensive focus on balancing participation, and not enough on ensuring access to knowledge and a clear enough method for assessing and arbitrating differences in views. Thus the "common ground" reached was both ambiguous and contested, and actions were marginal and divorced from critical thinking and analysis. The workplace collaborative research project was much more strongly based on strong empirical research. However, the process has still not been able to bring about the more difficult "upstream" changes in work organization, perhaps because of its restricted mandate and limited involvement from the full range of levels in the organizational hierarchy.

Serious challenges to carrying out action research on workplace health issues exist. Even in the best of times, significant methodological and practical difficulties exist: ensuring confidentiality, securing an adequate sample size, dealing with employee turnover, measuring complex constructs, and sorting out cause-effect relationships during ongoing change. At the present time, additional challenges stem from economic and political climates in most English-speaking industrialized countries identified in this book that are antagonistic to worker participation. Workplaces are becoming more fragmented, contingent and insecure as companies increasingly rely on contractors, temporary workers and off-site workers (Krahn, 1995). Such trends are associated with the increasing power of financial organizations outside of workplaces, above governments and with little interest in open, balanced discussion of worker health concerns.

In such circumstances, new kinds of workplace health concerns will continue to arise, requiring creative approaches to collaborative problem-solving by multiple stakeholders. Small steps in the direction of collective inquiry, research and action on worker health are being taken, as evidenced here. Action research can provide a promising basis for multi-stakeholder action and change to improve the health of workers.

7.6 KEY MESSAGES

- New approaches to research and action are needed to address today's ambiguous, contentious and multi causal workplace injuries and illnesses.

- The tradition of action research provides a promising basis for participatory, reflective and action-oriented strategies for researcher-workplace partnerships.

- A Future Search conference on repetitive strain injuries (RSI) in Ontario workplaces illustrates both the opportunities and pitfalls of a consensus-based multi-stakeholder visioning process in terms of securing diverse and balanced participation, finding common ground, and stimulating action.

- Collaborative research on RSI at a Toronto newspaper company succeeded in increasing workplace awareness, early reporting of symptoms and broader policy change but faced challenges in reorienting broader organizational practices which put employees at risk for RSI.

- Action research approaches appear to have the potential for generating broad participation in and commitment to actions to reduce the burden of RSI.

- Users of action research need to work hard to balance the sometimes competing goals of learning and scientific rigor, promoting consensus while accepting difference, and encouraging critical reflection without stifling action.

- Multi-stakeholder action approach to complex workplace health issues should seek to be both fair (i.e. open and inclusive) and competent.

7.7 REFERENCES

Beaton, D.E., Cole, D.C., Manno, M., Bombardier, C., Hogg-Johnson, S. and Shannon, H.S., 2000, Describing the burden of upper extremity musculoskeletal disorders in newspaper workers: What difference do case definitions make? *Journal of Occupational Rehabilitation*, **10**, pp. 39-53.

Brown, L.D., 1983, *Managing conflict at organizational interfaces.* (Reading, MA: Addison-Wesley).

Epp, K., 1986, *Achieving health for all: a framework for health promotion.* (Ottawa; Ministry of Supply and Services).

Evered, R. and Louis, M.R., 1981, Alternative perspectives in the organizational sciences: "Inquiry from the Inside" and "inquiry from the outside". *Academy of Management Review*, **6**, pp. 385-395.

Forester, J., 1999, *The deliberative practitioner: Encouraging participatory planning processes.* (Cambridge: MIT Press.)

Gustavsen, B., 1996, Development and the social sciences: an uneasy relationship. In *Beyond Theory: Changing Organizations through Participation,* edited by Toulmin, S. and Gustavsen, B., (Amsterdam/Philadelphia: John Benjamens).

Israel, B.A., Schurman, S.J. and Hugentobler, M.K, 1992, Conducting action research: relationships between organization members and researchers. *Journal of Applied Behavioral Science*, **28**, pp. 74-101.

Krahn, H., 1995, Non-standard work on the rise. *Perspectives on Labour and Income.* Winter: **35**.

Labonte, R., 1994, Death of a program, birth of a metaphor: the development of health promotion in Canada. In *Health promotion in Canada: Provincial, national and international perspectives*, edited by Pederson, A., O'Neill, M. and Rootman, (I Toronto: W.B. Saunders).

Mason, R.O., and Mitroff, I.I., 1981, *Challenging strategic planning assumptions: Theory, cases, and techniques,* (New York: John Wiley & Sons).

Mergler, D., 1987, Worker participation in occupational health research: Theory and practice. *International Journal of Health Services*, **17**, pp. 151-167.

Moore, S. and Garg, A., 1997, Participatory ergonomics in a red meat packing plant, part 1: evidence of long term effectiveness. *The American Industrial Hygiene Association,* **58**, pp. 127-131.

Neuman, G.A., Edwards, J.E. and Raju, N.S., 1989, Organizational development interventions: a meta-analysis of their effects on satisfaction and other attutudes. *Personnel Psychology*, **42**, pp. 461-489.

National Center for the Dissemination of Disability Research, July, 1996, *A review of the literature on dissemination and knowledge utilization.*

Occupational Safety and Health Administration (OSHA), Department of Labor, 2000, Ergonomics Program, Final Rule. *Federal Register,* **65**, pp. 68262-68870.

Pasmore, W. and Friedlander, F., 1982, An action-research program for increasing employee involvement in problem solving. *Administrative Science Quarterly,* **27**, pp. 343-362.

Polanyi, M., 2001, Towards common ground and action on repetitive strain injuries: An assessment of a Future Search conference. *Journal of Applied Behavioral Science,* **37**, pp. 465-487.

Polanyi, M.F.D., 2002, Communicative action in practice? Future Search and the pursuit of an open, reflective, and non-coercive large group change process. *Systems Research and Behavioral Science.* Planning and Design, **19** (4), pp. 357-366.

Polanyi, M., Cole, D.C., Beaton, D.E., Chung, J., Wells, R., Abdolell, M., Beech-Hawley, L., Ferrier, S.E., Mondloch, M.V., Shields, S.A., Smith, J.M. and Shannon, H.S., 1997, Upper limb work-related musculoskeletal disorders among newspaper employees: cross-sectional survey results. *American Journal of Industrial Medicine,* **32**, pp. 620-628.

Rantanen, J., Lehtinen, S., Kalimo, R., Nordman, H., Vainio, H. and Viikari-Juntura, E., 1994 New epidemics in occupational health. In *Proceedings of the International Symposium on New Epidemics in Occupational Health,* (Helsinki, Finland, May 16-19. Helsinki: Finnish Institute of Occupational Health).

Renn, O., Webler, T. and Wiedemann, P., 1995, *Fairness and competence in citizen participation: evaluating models for environmental discourse.* (Dordrecht: Kluwer Academic Publishers), pp. 35-86.

Schenk, C. and Anderson, J., 1995, *Re-shaping work: union responses to technological change.* Don Mills, Ont.: Ontario Federation of Labour, Technology Adjustment Program.

Webler, T., 1995, "Right" discourse in citizen participation: an evaluative yardstick. In *Fairness and competence in citizen participation: evaluating models for environmental discourse,* edited by O. Renn, O., Webler, T and Wiedemann, P., (Dordrecht: Kluwer Academic Publishers), pp.35-86.

Weisbord, M.R. and Janoff, S., 1995, *Future search: An action guide to finding common ground in organizations and communities,* (San Francisco: Berrett-Koehler).

Wells, R., Cole, D. and the Worksite Upper Extremity Research Group, 2001, Intervention in computer intense work. In *Prevention of Muscle Disorders in Computer Users: Scientific Basis and Recommendations. The 2nd PROCID (Prevention of muscle disorders in operation of Computer Input Devices) Symposium,* edited by Sandsjö, L., and Kadefors, R., 8-10 March 2001. (National Institute for Working Life/West, Göteborg, Sweden), pp. 199-125.

CHAPTER 8

Evidence-based Ergonomic Interventions in the Manufacturing Sector

Robert Norman

8.1 THE PROBLEM

In many businesses and industries, upper limb and low back disorders are, by far, the largest contributors to the reporting of work-related pain and injury and time lost from work. When a single precipitating overload incident occurs to account for these injuries, relationships between cause and effect are relatively obvious. Since the factor that caused the risk and resulted in the injury is clearly known, it may be possible to effectively intervene to remove the risk factor. Unfortunately, in the majority of contemporary occurrences, the pain cannot be attributed to a single incident; the pain just gets slowly worse until the worker cannot or will not cope with it any longer. The worker reports the pain and often goes off work because of disability. In yet other cases, the "incident" or even the nature of the work that has precipitated the pain report appears to skeptical observers to be so innocuous as to be incredible.

The absence of an obvious injury risk factor for slow onset low back or upper limb pain, often combined with the inability of the medical profession to specifically diagnose the cause of the pain (e.g. Nachemson, 1992), frequently results in skepticism about the legitimacy of the pain report. There is also often skepticism about the relationship of the reported pain to the physical demands of work (e.g. Hadler, 1990). This, in turn, leads to speculation and assumptions about what the real "risk factors" are for the reporting of low back and upper limb pain. Beliefs about risk factors are then acted upon in an effort to reduce costs. In the experience of this author over a period of many years, the most common speculations about causes of pain reports have to do with job dissatisfaction, plant politics, unsafe worker behavior or workers who are too unfit to meet the job requirements. Suspicions about "psychosocial" or "personal" factors have been around a long time. These beliefs then lead to worker behavior-based interventions such as back injury prevention education or wellness programs or claims challenging. I have rarely heard comments from management about the need for improved design or organization of work to make physical or cognitive demands manageable. Indeed, often comments are heard that "all of the heavy work has been engineered out and they still report injury".

Interventions that are based upon incorrect assumptions about risk factors for pain reporting are not only expensive they are ineffectual. Interventions must be aimed at reducing or eliminating scientifically validated causes of pain reports (known risk factors), not unsubstantiated beliefs or assumptions about what the risk factors are. Table 8.1 is a list of often heard and conflicting, assumed "causes" and related "interventions" to reduce and deal with low back and upper limb pain and injury reports. The reporting of pain by a worker is important, regardless of cause or medical diagnosis, because it is the penultimate step to work absence.

Table 8.1 Some assumed causes of pain reports and related interventions to reduce them.

Assumed cause	Intervention	Assumed cause	Intervention
Workers behave unsafely.	Provide education on safe work practices, monetary or substantial gift incentives for no lost time claims to encourage "safe work practices".	The inappropriate design of jobs, tools, materials, work space layouts prevent workers from "behaving safely" or using safe work practices they have been taught.	Design the work so that education that has been provided can be used.
Workers are out of shape.	Provide fitness facilities and wellness programs.	Work rates are too fast and durations, particularly with overtime, are too long; Design of the job, tool, material, work space layout are ill-conceived and unnecessarily tiring. Workers are too tired to participate in a wellness program.	Design work rates, including planned short recovery periods and micropauses, so that workers can keep up without undo fatigue; Hire enough workers so that production schedules can be met without excessive overtime.
Aging brings aches and pains. The workforce is aging.	The pain is not work related. Challenge the claim. Hire a younger workforce: pay off or lay off.	Physical demands at work have accelerated "aging".	Design the work to both minimize disability and accommodate an inevitably aging workforce.
Compensation payments are too generous; this encourages work absence and slow return to work.	Challenge the claim. Lobby to change fee schedule. Keep in constant contact with the injured worker at home. Keep injured workers at the workplace even if there is no useful work for them.	Work rates, durations and other features of the design of work are too demanding, particularly for a worker with an impairment. The number of workers is declining but production volume is increasing. We are working 55 second minutes. Five seconds of recovery is not enough.	Provide well-designed and meaningful alternative or modified work so disability is minimized. Design work so physical demands are sufficiently low so even workers with impairments can perform. This will allow the return of the injured worker, minimize risk of a first injury.
Workers of today just will not tolerate the job demands like they used to.	Provide high salaries, additional monetary or substantial gift incentives for no lost time claims to encourage "safe work practices" and keep workers at work.	Work rates, durations and other features of the design of work are too demanding for any worker.	Design work so the physical demands are sufficiently low so that even workers with impairments can do them. This will allow return of injured worker and also minimize risk of a first injury.

8.1.1   Known Risk Factors (causes) for the Reporting of Pain at Work

Employers should pay attention to the reporting of pain, even in the absence of a medical diagnosis, because a pain report is the first step taken by an employee toward time lost from work. Accurate medical diagnosis is important for specifying medical treatment and clinical rehabilitation but it is not important for informing work site interventions to prevent injury to reduce disability resulting from injury, thereby facilitating retention or early and sustainable return to work. Worker perceptions of pain regardless of medical cause are extremely important. In some businesses, once an employee is absent because of pain, it becomes very difficult for some to return to work. Long-term disability is expensive. Chapter 1 by Kerr and Norman reviews risk factors in detail. Highlights here, to put this ergonomic intervention chapter in context, may be helpful. There have been speculations for many years about the importance of several categories of these factors. Only recently have relative contributions of a variety of variables to pain reporting become known with some certainty.

Figure 8.1 is a conceptualization of many hypothesized risk factors for the reporting of pain or tissue damage proposed by researchers from a wide variety of disciplines. The many possible variables are organized into three logical sets, psychosocial (not psychiatric) variables, physical variables, and personal variables. Physical variables are related to the biomechanical and physiological demands of tasks and jobs either forced by the engineering design of the work and work schedules or by work methods chosen by the worker. Psychosocial variables are related to workers' observations and perceptions about their work, workplace and other people, accurate or not. Personal variables are inherited or acquired characteristics, practices and behaviors of individual workers and management. Not all of variables in Figure 8.1, or beliefs alluded to above, have support from research literature. An extensive epidemiological study of auto assembly and assembly support workers revealed a smaller subset of independent risk factors for reporting of low back pain than proposed in Figure 8.1 (Kerr *et al.*, 2001). This subset is shown in Figure 8.2.

This study clearly showed the two most important categories of variables are biomechanical and psychosocial; the least important was personal variables. Collectively, psychosocial variables accounted for approximately 12% of the total variance in prediction of the reporting of low back pain. Physical demands of the work accounted for approximately 30% when workers' numerical ratings of physical exertion on the job were included. Personal variables accounted for less than 5%. There is no reason to believe variables would be substantially different for predicting injury to upper limbs or other body parts. We now know the relative contribution to the reporting of pain at work and how to reliably measure all of the variables. Unfortunately, this study also showed there is no single "magic bullet" risk factor that, if removed, will significantly reduce work-related disability. No single variable by itself accounted for a great deal of pain reporting but all variables combined accounted for nearly 50% of pain reporting. This knowledge is an appreciable advantage in designing an effective intervention program but shows no single intervention – such as a wellness program, more supportive workplace environment, or better tool design – is likely to prevent injury on its own, rather, coordinated efforts are required.

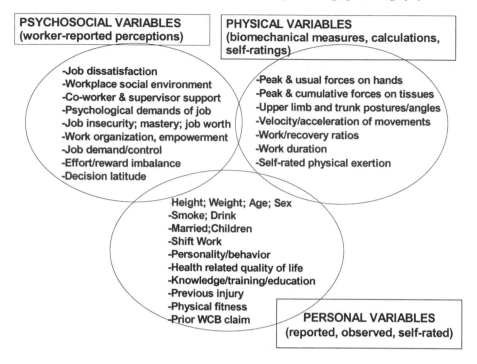

Figure 8.1 Categories of hypothesized risk factors for the reporting of work-related pain. Some factors have been proven risk factors, many have not.

8.1.2 Principles of Ergonomic Intervention

Ergonomics is the practice and science of "human-centred" design of work. This is to be compared with product or service or profitability or manufacturing process-centered design. More specifically, ergonomics is the process of designing, modifying or organizing tools, materials, equipment, work spaces, tasks, jobs, products, systems and environments to match psychological, social, anatomical, biomechanical and physiological abilities, needs and limitations of people. Thus, the scope of ergonomics involves not only physical and engineering considerations but also perceptual, cognitive, social and organizational aspects of work. Moreover, ergonomics includes interventions aimed at improving work at both the level of individuals (micro ergonomics) and at the level of work organization (macro ergonomics).

The objectives of ergonomics are to design work to improve productivity and the quality of the product or service provided by the company without compromising employee health, safety or the quality of working life. Ergonomics focuses on people in the system of the production process, maintenance and use of goods or services. High quality, productivity, profitability and safety are consequences of work that is designed with all of the abilities, needs and limitations of humans in the system considered at the time of the design of the work, product or service, not afterwards.

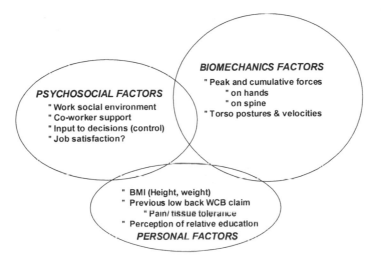

Figure 8.2. Statistically independent risk factors for the reporting of low back pain in a large auto assembly plant (Kerr *et al.*, 2001). BMI is body mass index, a weighted ratio of height to weight. The ovals are different sizes to roughly (not to scale) reflect differences in proportions of total variance accounted for by each category of variables.

Forcing humans to adapt to poor workplace, product or service designs after the design stage is complete has adverse monetary and social outcomes. All employers have long recognized that purchasers of their products or services demand compatibility with their needs and capabilities. It appears that few recognized that quality, productivity, profitability, health and safety are, similarly, related to the abilities, limitations and needs of their workforce. Compensation costs are part of the evidence for this statement and these costs are resulting in increased interest in interventions that work. All interventions should be aimed at reducing or eliminating proven, not imagined, risk factors. Ideally, the risk factors that account for the largest amount of variance should be reduced first and the variables with less impact, later. The problem is that many risk factors affect the reporting of pain and none, by itself, accounts for a large part of the variability in predicting pain reporting. Therefore, intervening on any one of these variables is futile if cost reduction, by the prevention of initial injuries, reduction of disability or safe and sustained return to work are objectives. No single risk factor accounts for an appreciable proportion of variance. Together, they account for a lot. Although statistically independent risk factors are, by definition, not highly correlated with each other, it is possible to identify categories of variables that, if changed, may have positive effects on other variables not directly attacked. Variables of this type should be near the top of the list for intervention. Practicality of intervening is, of course, also a major consideration. Since a formal or informal business rationale will probably be required, particularly if the intervention is expensive, it is preferable to intervene on variables that can be measured. This will allow assessment of progress towards improvement, as measured by the lowering of a risk factor immediately, before downstream objectives such as reductions in pain and injury reports or absenteeism costs or productivity or quality improvements will have an opportunity to be seen, assuming the intervention is successful.

Even with these intervention principles and knowledge and the awareness of the need to intervene in a way that simultaneously affects as many variables as possible, there is sufficient interdependence of some of these variables to present serious questions about where and how to start to intervene. Should one start by strategically attempting to modify psychosocial factors, physical factors or personal factors?

We advocate starting with the biomechanics factors for several reasons. The physical risk factors account for the largest proportion of variance, therefore one might decide to begin by improving the design of work. For example, altering the design of a work space layout so that the amount of forward torso bend is reduced improves trunk posture by straightening the worker up. This also reduces trunk velocities because the range of trunk motion is reduced. Simultaneously, sizes of peak and cumulative forces on the spine are reduced. All these variables are important and proven risk factors and a single intervention can affect several at once. Improving job rotation schedules reduces cumulative forces on hands and spinal tissues but does not reduce peak forces. In fact, a rotation may expose people not exposed when there was no rotation. Conscious attention to work rates and durations can improve work/recovery ratios and also influences the sizes of peak and cumulative forces on hands, arms and spine.

One might expect additional risk reduction payoff by making physical changes. This type of change is highly visible to the workforce and will be seen to be demonstrations of actions by management that will have positive effects on a number of psychosocial variables such as perceptions about work place environment and job satisfaction. If the process of making these changes actively involves the workers, because they have valuable knowledge about their jobs to contribute to redesign or initial design solutions, observations and perceptions of job control and, perhaps, demand should be improved. Moreover, if ergonomics education is part of the intervention, a personal variable, physical and organizational improvements in the design of work should allow more effective use of education about safe work practices. Of course, another option is to start with psychosocial variables, perhaps by improving perceptions of workers about the work place environment and attitudes of management about their health and safety or the control they have over their work. A third option, unfortunately, one often used as a sole intervention effort, is to work first on personal variables such as promoting improvements in fitness or wellness of the workforce or in their knowledge about injury by means of training courses on "correct" lifting.

Our experience in delivering back education courses to several thousand workers and supervisors steers us away from education as a sole intervention for musculoskeletal disorders. The research literature generally does not show large cost reductions attributable to back education or fitness programs in the absence of job design improvements. In addition, we have anecdotal evidence from many of the workers and some systematically collected evidence, that this type of personal intervention and, indeed, psychosocial interventions, are seen to be disingenuous actions by management in the absence of improvements in job design. The issues of where and how to intervene, even knowing the risk factors, is complex. Several literature reviews on the topic of ergonomic intervention are available (e.g. NIOSH, 1997; Norman and Wells, 2000). It is necessary to try to sort the interactions out by conceptualizing how the workplace works so that the intervention strategy is not only evidence-based but it is also practical. A mental model for intervention is essential.

8.1.3 A Proposed Model for Ergonomics Intervention

Figure 8.3 is a conceptualization or model both of risk factors that act on individual workers, the inner three circles, and effects on these risk factors initiated or controlled at the level of the corporation or plant, the underlying plant or corporate "culture", a fourth circle. We would argue that this model of intervention is related, not only to injury risk, but also to quality, productivity and profitability of the goods or services that a company produces. Management buy-in to intervention is critical to the success of the intervention. Corporate beliefs, values and culture are highly related to how work is designed and organized and management decisions affect what happens on the plant floor. The research evidence supports the risk factors that operate at the level of the individual worker. The nature of the impact of corporate culture, although highly suspected to be critical, is less certain.

The model proposes that Corporate Organization and Culture Factors provide an underlay and interact with the three other, mutually interacting categories of factors, Psychosocial Factors, Work Design Factors, and Personal Factors discussed earlier. The psychosocial factors are largely perceptions of the workforce about their work, their colleagues, their supervisors and others in management and a variety of other aspects of their environment. They are not psychiatric factors. Work design factors include biomechanical, physiological, perceptual, cognitive and engineering aspects of the work, including work duration, rate and related work/recovery cycles. Personal factors are the inherited and acquired biophysical and psychological traits and behaviors that people bring to their work, including knowledge and experience.

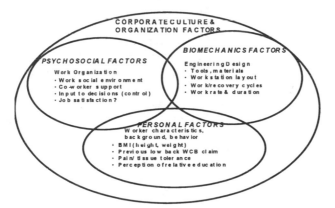

Figure 8.3 A proposed, four component ergonomics intervention model we believe is relevant, to reducing risk of injury, and with issues of product or service quality, productivity and profitability.

On a macro level, the organization of the corporation or plant and the corporate or plant culture are influenced by perceptions of management. These include factors such as perception of how work must be done within the legislative framework of the country or province, how to profitably compete in a complex global economy, how to deal with a reducing and aging workforce, how to exploit technology and the changing nature of work. These perceptions lead to corporate or plant beliefs that often become values. The values, in turn, influence action by corporate and plant management. Examples of actions at this level are policies, production rates, product volumes and schedules. Priorities among quality, productivity, profitability, health, safety and quality of working life, practices regarding accommodation of injured workers, participation by labor in decision making, the size of the workforce, overtime, are yet other examples of management action that, in the perceptions of the workforce, reflect management attitudes. Therefore, decisions at the macro level of the model most certainly affect many risk factors at the micro level.

On a micro level, the level of individuals within the corporation, the three interacting categories of variables, informed by the risk factor research, are again identified. If one considers personnel selection as a type of intervention, all of these variables are modifiable.

The multi-factor nature of the intervention problem indicates that for effective intervention, the organization has to be considered as a whole, with intervention at both levels, macro and micro, and in all categories of risk factors before costs will decline. Intervention at only one place in the system will not account for sufficient variance to show a difference in prevention of initial injury, recurrence, or disability, and almost certainly not on the costs of injury-related work absence.

For example, public statements of commitment by senior management to reducing injury incidence and severity is absolutely necessary for there to be any hope of improvements in injury records. But, in the absence of supporting behavior by managers such as by spending money on risk-reducing job design changes at the level of the individual worker, the most clearly proven risk factors, adverse postures and excessive cumulative and peak forces on body tissues, will not be reduced and pain reports and injuries will continue. Working solely on personal factors, such as education, stress management, physical fitness, will not be effective either because the physical risk presented by the line rate, tool design, prolonged adverse postures or repetitions will not have been reduced. Moreover, negative perceptions by the workforce about lack of company concern and a number of other proven psychosocial risk factors will persist or escalate. Indeed, we speculate that if education is the sole intervention, perceptions of management commitment to their workforce by workers could be substantially improved by the presence of very senior managers in ergonomics education courses, side by side with the workers and other personnel.

Unfortunately, there is very little research on interventions that is sufficiently well designed and implemented to be unequivocal in its conclusions and convincing to skeptical managers. This type of research is very difficult to do for many reasons that include: a work environment that changes more rapidly than the time it takes to complete a properly measured study with appropriate comparison groups; the difficulty in finding, and ethics and natural change in the work place that inhibits maintenance of proper comparison groups for duration of a study; high costs of studying sufficiently large groups to obtain acceptable statistical power to be confident that lack of observed differences between intervention and comparison groups are real; convincing companies to get involved in this type of research given inevitable interruptions and risk of finding results perceived by management or labor to be adverse or costly. Fortunately, we have been able to begin to test the intervention model proposed.

8.1.4 An Evidence-based Ergonomics Intervention Example

This ongoing project is an extension of a six-year Ergonomics Initiative with an automotive parts producer and the Canadian Autoworkers' Union. The first several years were devoted to the development of methods for quantifying worker exposure to physical demands of tasks, identification of physical, psychosocial and personal risk factors for the reporting of work-related pain, and ergonomics education of workers and salaried employees including management. Once the risk factors and methods of measuring them were identified, the challenge became working with labor and management to develop and evaluate an ergonomics intervention program that capitalized on this knowledge.

The goal of the intervention project was to minimize injury risk while maintaining or improving quality and productivity in the work area. The participating company is a multinational of medium size that makes molded foam pieces such as seat pans, dashboard impact absorbers, arm and head rests mainly for the automotive industry. There are several plants in Canada and in other countries of approximately 200 or 300 employees each. Some work involves production lines where workers remove parts from a conveyor that moves molds at a constant speed. They then have to scrape the molds clean and wax them to prepare them for the next chemical pour that then becomes foam. Off the line workers manually trim unwanted flashing from the pieces, inspect, pack and do other tasks in preparation for shipping. There is a job rotation procedure in place that allows most workers to move from station to station to vary the work. Low back and upper limb pain are the most frequent problems.

Intervention Strategy

The previous research showed that it is necessary to attempt to reduce as many risk factors as possible in all three categories described in Figure 8.3, psychosocial, biomechanical and personal. The model identifies the importance of also engaging senior administrators as part of the ergonomic intervention process. We consciously decided to attack the biomechanical risk factor category in a direct attempt to reduce physical risk factors by improving job design. We hypothesized that there would also be an indirect psychosocial benefit of improvements in job satisfaction if visible physical changes were implemented – after appropriate consultation with plant and corporate stakeholders. Therefore, to also directly address the "psychosocial risk factors" of worker control and input to decisions, in addition to capitalizing on their knowledge about their jobs and working conditions, we elected to use a "participatory ergonomics" as the core of the intervention process. Kourinka (1997) has defined participatory ergonomics as practical ergonomics with participation of the necessary actors in problem solving. The participation is not limited to workers but, rather, includes all levels of the hierarchy with first-hand experience about the problems at hand. To address "personal risk factors" and so that ergonomics participation could be better informed, we provided ergonomics education for a number of the participants.

Description of Intervention

The participatory change process began with the formation of an active Steering Committee (SC) and an Ergonomics Change Team (ECT). The SC comprised corporate management, plant management, union representation, and research team (RT) members. The objectives of implementing this committee were to develop and maintain corporate and plant management interest and input as the intervention program evolved, to foster regular communication between these senior levels of management and plant labor personnel, and to solve administrative problems as they arose. The SC met regularly by telephone conference. The RT provided suggestions to the plant and corporate management concerning the kinds of roles that might be represented on the ECT. The company selected ten members to serve on the ECT. They included corporate management, the union, lead hands, human resources, engineering, managers from production and tooling and maintenance and, quite often, the plant manager himself. Two production lines were involved, an intervention line and a comparison line that was promised intervention at a later date.

The ECT was the ergonomics action committee. They received ergonomics education about risk factor identification, measurement and management; they oversaw the priority setting for change in workstations, tools and production and interacted on the plant floor to solicit input and explain changes to workers. Indeed, the hope and expectation was that sustainability of the initiative after the RT withdrew would be in the hands of the ECT. The ECT was self-directed and worked by consensus.

Outcome and Evaluation of the Intervention Products and Process

The ECT initiated 21 ergonomic actions. All changes resulted in improvements, on an individual level, in muscle electrical activity (EMG) in the upper limbs, or spinal loading (computerized biomechanical model, 4DWATBAK), or trunk or limb posture (judgment of RT ergonomic experts or by the biomechanical software), in preference ratings or in general comments by workers. The effectiveness of the participatory process was formally evaluated (NUDIST) from transcripts of scripted interviews with corporate and plant management, the ECT and floor workers. This qualitative analysis revealed general satisfaction with the process of ergonomics change, although both labor and management personnel who had been directly involved on the ECT or SC were more positive about communication successes than those not directly involved. Interestingly, these improvements in perceptions did not reflect group (compared with individual case study) data on physical loading attributable to the job design changes.

There were no measurable group level reductions in physical demand although ratings of pain increased less on the intervention line than on the comparison line. The reason may have been that the ECT was not allowed to change the conveyance system, only tools, seating, materials tables and other small features of the environment. Conveyance system changes are very expensive even though its design is felt to be responsible for worker-identified pain because of adverse trunk and arm postures that the device imposes. It should be noted that a modified (tilted) conveyance system has now been installed in a different plant and its efficacy is being evaluated by part of the RT. A further limitation was that it proved to be difficult to maintain an experimentally uncontaminated comparison line. Workers on that line adopted several of the ergonomics changes that were intended, temporarily, only for the intervention line before a post-intervention analysis was conducted.

Conclusions about Evidence-based Ergonomics Intervention

The study showed that an evidence-based, intervention model assisted, participatory process can have positive outcomes on perceptions of ergonomic action in the absence of reductions of physical demand. Whether these perceptions and the intervention process are sustainable under these conditions remains to be seen. We suspect that they are not and will evaluate this in a new study at another plant and a reassessment of this plant. The study also confirmed general knowledge by ergonomists that communication about ergonomics change with the workers, all of whom have pride and ownership in their jobs, is essential before changes are brought to their jobs, even demonstrably beneficial changes. Indeed, a consultative and inclusive process of ergonomics change is probably as important as the change itself to the utilization of an intervention. Good ideas about job design often fail if the process of change is flawed.

The study also showed that standard textbook epidemiological experimental design procedures are very difficult to follow in a work site intervention of this type. Corporate management was not interested in randomization. They knew what plant they wanted to be part of the study. It took considerable persuading to convince them of the need for a control or comparison group but once they agreed, within the plant, the two lines that we were allowed to engage were decided by the plant managers, they were not randomized. Even the comparison line was proscribed and it proved impossible to withhold ergonomic interventions from this line. There was an understandable impatience of some corporate and plant management and workers with the time it took to begin to intervene because of the time it took to analyze baseline data. In a subsequent study we have decided to intervene as we go and to gradually build the database to obviate this problem. Large-scale group analyses on work site interventions that involve job modifications are very difficult if not impossible. Intervening on production processes is the domain of plant and corporate personnel and outsiders are not suffered gladly by some. On the other hand, we were able to conduct a very large case-control study in a production facility with relatively little problem (Kerr *et al.*, 2001; Norman *et al.*, 1998) but there we were looking for risk factors, the domain of researchers, and not interfering with production significantly.

A final outcome of this study, at the request of the company, was development of an "Ergonomics Program Implementation Blueprint" (Wells *et al.*, 2000). This is a document that describes, in detail, a step by step ergonomics intervention program and process, including a number of risk factor measurement methods. We are in the process of evaluating the coherency and usability of the "blueprint" in another plant of the same company. Documents like this are highly desired by workplace partners so something is left for them to continue to use after the research team has withdrawn.

8.2 SUMMARY

Preventing injury or at least minimizing disability if injury occurs, while at the same time safely maintaining production schedules, is a complex business. If this were easy, the social, compensation and production costs of work-related low back and upper limb pain would be much lower than they are. Informed employers would have designed safe work.

8.3 KEY MESSAGES

- There are several groups of interacting risk factors that account for almost 50% of the variability in the prediction of who will report work-related pain. These groups can be categorized as Psychosocial Factors (worker perceptions), Physical Demand Factors (biomechanical variables), and Personal Factors (inherited or acquired variables).

- Statistically, the most important category is Physical Demand related to job design. This is the category that affects the sizes of peak and cumulative forces on body, forces on the hands and trunk and upper limb postures, all of which have been proven to be very strong risk factors. One recent finding is the epidemiological evidence that cumulative forces, as a result of high repetitions or prolonged work over the course of a shift, even if the peak forces in a cycle of work are low, are independent risk factors. This risk factor is in addition to the risk imposed by high peak forces.

- The second most important category is Psychosocial. What is and what is perceived to be are not always the same thing. But perceptions are extremely important and have been proven to be quite strong risk factors. Perceptions by workers are very much influenced by observation of company action or inaction.

- The least important category is Personal Factors. Interestingly, many dollars are spent on trying to change this type of factor, for example by short educational courses, without first attempting to improve job design. Worker

---

and management education is important but it does not help to educate a worker about low-risk load handling principles and then send them back to a job design that does not allow them to use the principles.

- Work site interventions designed to reduce the costs of work-related injury, pain and absence must be aimed at simultaneously reducing or eliminating as many proven risk factors as possible, particularly the Physical Demand and Psychosocial Factors. It is not helpful to work on only one or two known risk factors and expect to see measurable reductions in no lost time or lost time injuries or reductions in injury-related absenteeism or dollar cost savings. No single variable accounts for enough of the total cost to make a difference by itself.

- We propose that intervention should start by improving job design, not personal factors or perceptions and that a participatory process of ergonomics change between management will have both direct and indirect positive influence. However, previous history of relations between management and labor sometimes reduces the willingness of either or both parties to participate for various reasons. Some companies are farther along the road to effective participation than others.

- If a participatory process can be effectively adopted, ergonomics education about injury mechanisms, risk factors and how to measure them is necessary to ensure that discussions and decisions about intervention is of high quality. Education of workers on lower risk behavior should not be a substitute for improved job design.

- The process of ergonomics intervention is probably as important as the actual intervention. Implementing interventions without prior consultations with the workers who have to use them often results in the rejection of even good interventions. Communication among all stakeholders is critical to success.

8.4 ACKNOWLEDGEMENTS

Although the ideas presented in this chapter are those of the author, for which he takes full responsibility, they have evolved as a result of hours of conversation and the results of ergonomics laboratory and field research with several colleagues. He acknowledges the influences on his thinking of his colleagues at the Institute for Work & Health (Toronto), Doctors Donald Cole, Harry Shannon, John Frank, Michael Kerr and Terrence Sullivan. In particular, he is indebted to Doctors Richard Wells and Mardon Frazer and research assistants Andrew Laing and Patrick Neumann (now in Sweden) from the University of Waterloo. The ergonomic intervention described at the end of the chapter is a very large multidisciplinary project that includes very important contributions from most of the people named above, Doctors Nancy Theberge, Lawrence Brawley and Robert Kerton of the University of Waterloo, and corporate and plant management, workers and worker representatives from the Woodbridge Group and the Canadian Auto Workers' Union. The author is indebted to them all. The much appreciated funding for the research upon which this chapter has been based has come from HealNet, the Workplace Safety and Insurance Board of Ontario, The Institute for Work & Health, General Motors of Canada Ltd, A.G. Simpson Ltd, and the Woodbridge Group.

8.5 REFERENCES

Hadler, N.M., 1990, Cumulative trauma disorders: an iatrogenic concept. *Journal of Occupational Medicine,* **32**, pp. 18-41.

Kerr, M.S., Frank, J.W., Shannon, H.S., Norman, R.W., Wells, R.P., Neumann, W.P. and Bombardier, C., 2001, Biomechanical and psychosocial risk factors for low back pain at work. *American Journal of Public Health,* **91,** pp.1-7.

Kourinka, I., 1997, Tools and means of implementing participatory ergonomics. *International Journal of Industrial Ergonomics*, **19**, pp. 267-270.

Nachemson, A.L., 1992, Newest knowledge in low back pain: a critical look. *Clinical Orthopaedics and Related Research*, **279**, pp. 8-20.

National Institute for Occupational Safety and Health (NIOSH), 1997, *Musculoskeletal disorders and workplace factors: A critical review of epidemiologic evidence for work-related musculoskeletal disorders of the neck, upper extremity, and low back. Cincinnati*: US Department of Health and Human Services.

Norman, R., Wells, R., Neumann, P., Frank, J., Shannon, H. and Kerr, M., 1998, A comparison of peak vs. cumulative physical work exposure risk factors for the reporting of low back pain in the automotive industry. *Clinical Biomechanics.* **13,** pp. 561-573.

Norman, R. and Wells, R., 2000, Ergonomic interventions for reducing musculoskeletal disorders. In: Sullivan, T. (Ed.). *Injury and the New World of Work,* (Vancouver: UBC Press), pp. 115-139.

Wells, R., Norman, R., Frazer, M. and Laing, A., 2000, *Ergonomics Program Implementation Blueprint. Ergonomics and Safety Consulting Services, Faculty of Applied Health Sciences,* University of Waterloo. (61 pages).

CHAPTER 9

Preventing Work-related Disability: Lessons from Washington

Robert D. Mootz, Gary M. Franklin and Thomas M. Wickizer

9.1 HISTORY, CONCEPTS, AND CHALLENGES OF WORKERS' COMPENSATION

Recognition that injuries in the workplace are a community problem, rather than just an individual issue, is not new. As early as 1713, Ramazzini documented dangers of work-related health problems in a work entitled *The Diseases of Workers* (Rothstein, 1990). Beals (1984) identified one of the earliest disability indemnity plans of the modern era among the society of seafaring pirates operating in the Western Hemisphere. Specific levels of compensation were delineated for eyes or limbs lost in the course of duty while plundering about the Caribbean. Throughout the 1800s awareness increased regarding relationships between occupations and certain injuries. The absence of a judicial recourse prompted calls for regulatory remedies that have fluctuated between worker and employer interests (Cheal, 1986). Early in the nineteenth century, precedent and case law seemed to blame employers for work injury. "The act of the servant" was considered to be "the act of the master" prompting employers to seek regulatory solutions. The pendulum swung the other way with the English Fellow Servant Doctrine in 1837 that protected the employer from liability if an injury resulted from a worker's or co-worker's negligence (Larson, 1952).

In the United States, social beliefs that employer liability was counter to the general economic good of modern industrialized societies prevailed well into the 1880s. In the 1842 legal decision of Farwell v. the Boston & Worcester Railroad, the employer was held immune from liability for injuries that resulted from the negligence of the company's switchmen (Larson, 1992). The state of Georgia abrogated the fellow servant defense in 1855 for railroads operating within their jurisdiction. By the 1880s England rectified some of their problems with an employer liability act, which was readily circumvented by individual worker contracts. Near the beginning of the twentieth century, the US Federal Employers Liability Act embodied many of the protections balancing between worker and employer interests that were being established by individual states.

Germany is credited with establishing the first formal governmental workers' compensation system in the later 1800s (Larsen, 1992). England followed suit shortly thereafter. Massachusetts was the first American state to establish a workers' compensation system in 1904. Washington State was among the earliest US jurisdictions to enact workers' compensation legislation in 1911 and was the first state to make such coverage mandatory (WSDLI, 1999b). Mississippi was the last US state to enact legislation in 1949 (Larson, 1992). By 1955, three-quarters of the US workforce was covered (Skolick, 1962). In 1970, the National Commission on State Workmen's Compensation Laws with representation from business, labor, workers' compensation agencies, insurance carriers, medical professionals, and academics was established by the US Congress to recommend a set of minimum standards for every state. Characteristics of contemporary workers compensation laws include:

- compulsory coverage
- no exemptions to coverage
- inclusion of domestic and agricultural workers
- full coverage of work-related diseases
- full medical and rehabilitation
- no arbitrary limits on duration or total benefits
- reasonable weekly time loss benefits.

9.2 CONTEMPORARY CHALLENGES

Workers' compensation systems are complex and varied around the world. They typically are set up as no-fault programs that aim to remove uncertainty regarding liability for work-related conditions for employers and workers alike. Although many of the issues are centuries old, formal workers' compensation systems are still quite young and they continue to face many of the same challenges of balancing worker, employer, and community interests.

Noble objectives notwithstanding, the evolution of modern systems is not without problems. Employers' Workers' Compensation costs in recent decades have increased from just over $2 billion in the 1960s to more than $62 billion in the 1990s (Pound, 1994). In the United States, the rise in medical expense per injury increased an average of 14% per year during the 1980s compared to an 8% rise in the medical component of the Consumer Price Index (Pound, 1994). Many systems have responded by introducing administrative oversight and cost containment strategies such as benefit caps, managed care, utilization review, and/or pre-authorization procedures into benefits administration.

Some of the biggest contemporary challenges include resolving problems that arise from the interface of regulatory mandates built on political compromises between business and labor with what might be considered straightforward clinical management issues. Balancing obligations of respective parties with incentives and protections can also be delicate. Protecting employers from liability for occupational injury or illness may serve as an incentive to ignore workplace safety issues. Incentives for compensating injured workers not working, especially when one is employed in an undesirable occupation or setting, may limit the motivation of workers to return to work. Doctors can often be placed in roles dictated more by statutory or adjudicative requirements and employer-employee relations than by clinical need (Fordyce, 1995). Thus, developing and implementing innovative solutions can be fraught with legal and political complexities.

9.3 THE WASHINGTON STATE APPROACH

In Washington, the Department of Labor and Industries (DLI) serves as both regulator and insurer for two-thirds of the state's labor force (about 1.7 million workers) employed by approximately 150,000 companies. The remainder are employed by 400 self-insured companies whose workers' compensation programs are certified and regulated by DLI. Although self-insured employers manage their own claims, they are required to follow the same laws and regulations as the State Fund. Approximately 180,000 claims are filed with the department annually with roughly 85% being accepted as occupational injuries or diseases (WSDLI, 1999a).

Among the legislated characteristics of the Washington State system are: free choice of provider by workers, no directing of care by employers, and accident reporting that comes primarily at the worker's initiative, rather than the employer's. As a result, work injuries and exposures usually first come to the attention of the department when a worker seeks medical attention and a doctor files an accident report. Although employers typically have internal policies regarding reporting of on-the-job injuries, employers in Washington are required only to file reports with the state when work illnesses or injuries result in hospitalization of two or more workers, fatality, or probable fatality. Workers' compensation claims to the state are initiated entirely by workers. Time-loss disability payments by the state begin only after four days of lost time. Other characteristics of the Washington State system are unique. Industrial insurance and the state's safety and health division exist within the same agency. Advisory committees (including business-labor, medical, and chiropractic committees) are established by law. The agency also has a formal research, evaluation, and training relationship established with the University of Washington. Access to fairly comprehensive population-based claims and medical billing databases also exists.

Although these characteristics are somewhat unique, the availability of consolidated staff and data resources, and the pivotal advice and consent role of business, labor and providers in Washington fostered an environment that prioritized research and extensive collaboration with provider communities. The opportunity to obtain information and work closely with physicians in the community yielded experience and innovations that may not be readily obtainable within other systems. Nonetheless, the experience gained in Washington has a great deal of relevance and applicability to other workers' compensation settings.

9.4 WASHINGTON'S POLICY ENVIRONMENT

A number of key factors have facilitated initiation of research studies conducted by the department, particularly policy and field studies. These factors include: 1) a formal statutory, advisory function for business and labor leaders with DLI; 2) a formal regulatory advisory function of separate medical and chiropractic committees representing the state-wide medical and chiropractic organizations; 3) a close collaborative partnership between the DLI and University of Washington health services researchers, based in statute; 4) access to population-based data systems; and 5) the state's regulatory environment (Wickizer *et al.*, 2001). The DLI's relations with the business, labor and practitioner communities have helped to ensure broad support for policy studies. In addition, collaborative partnerships with the researchers at the University of Washington help assure rigorous and scientifically sound design to studies and their implementation. The DLI's extensive data systems allow researchers access to detailed utilization and cost information on a population basis.

The regulatory environment has allowed the DLI to conduct innovative policy studies, once even permitting the pilot-testing of an entirely different reimbursement approach. Through purposeful, systematic pilot-testing of system changes, gleaning experience from them, then using the results to inform future tests and policy refinements, the department has been able to avoid instituting sweeping, system-wide changes prior to understanding their overall impact. This kind of approach helps foster innovation. One recent Stanford business study attributed achievements of the twentieth century's most successful and visionary corporations in large measure to innovation, evaluation, and testing regardless of whether or not a given initiative itself was successful (Collins, 1997). The key to sustained success was related to "experimenting" with innovations and then applying what was learned to future efforts more than any strategic planning efforts. Characteristics of this sort of approach have been deliberately used by the DLI in developing health policy initiatives.

9.4.1 Stakeholder and Constituent Partnerships

Perhaps because of the large size of the Washington system and its direct accountability as a public agency, formal organization and coordination with the many constituencies within workers' compensation has been a necessity. In general, the employer and labor communities are considered the principle beneficiaries and owners of the Washington system. Both employers and workers contribute to workers' compensation insurance premiums that directly fund both benefits and the bulk of administrative costs without using any general tax revenues. As a result, input processes from business and labor have been formalized through the establishment of advisory committees to both the industrial insurance and safety and health divisions of the DLI.

Such committees serve as two-way communication links between the DLI and the community. The term "stakeholder" has seen increased use to describe those individuals and organizations that may be impacted by policies and activities of government agencies or other large organizations. Incorporating input from representative stakeholders that may be affected by policies and decisions early in the policy development process fosters a more collaborative environment that reduces future disruptions and adversity. Although the specific circumstances of our regulatory structure helped dictate establishment of our advisory committees, the approach to organization and explicit processes that have been developed to secure input and maintain communication may have utility in any system. Several useful lessons from this approach to stakeholder management include:

- involving stakeholders in policy innovations early on to facilitate ownership of new ideas
- using divergent thinking early and incorporating more convergent thinking over time
- openly identify up-front the constraints under which any groups at the table might be working
- Advisory Committee structures that facilitate two-way communication
- Employer/Labor Advisory Committees that help identify important community issues that regulators or providers might not see
- Provider Advisory Committees that can foster constructive relationships as well as the ownership of new ideas.

9.4.2 Workers' Compensation Advisory Committee

The state's industrial insurance laws establish a Workers' Compensation Advisory Committee (WCAC) chaired by a representative from the department and include four representatives each from labor and business. In addition, the chair of the state's separate independent appeals agency (the Board of Industrial Insurance Appeals) is a member. This group is statutorily charged with "conducting study of any aspects of workers' compensation as the committee shall determine require their consideration". The WCAC serves as a resource to the system, representing the interests of the beneficiaries (labor and business). In recent years, the WCAC has been instrumental in initiating and supporting pilot research studies on long-term disability, managed care, and increasing occupational health services' expertise. Such pilot studies allow the testing of innovations that yield information that can assist in future system-wide improvements and policy development. The WCAC also establishes sub-committees that focus on claims administration, healthcare and other areas of interest. This mechanism helps assure that a forum for agency, employer, and labor perspectives exists throughout policy development and larger scale field projects.

9.4.3 Medical and Chiropractic Advisory Committees

Attending doctors who directly care for injured workers are in the best position to readily identify clinical problems as well as workplace and administrative problems that may increase risk of long-term disability (Mootz, 1999). Two provider advisory committees have been established by administrative rule with members nominated by the state's medical and chiropractic associations. Committee members bring direct patient care experience to the table and a substantial number of members on both committees have significant occupational health experience. Provider committees meet monthly with department staff to focus on quality of care issues, policy development, claims administration and reimbursement, guidelines development, among other things, and serve as a resource for communicating with community providers in the state.

Our experience suggests that policy and guidelines that have been developed with extensive involvement from community-based physicians are much more likely to be meaningfully understood and more readily adopted. Examples of projects that have been developed with the provider communities include treatment guidelines (WSDLI, 1999a), provider training materials (WSDLI, 1999b; WSDLI, 1999c), state-wide basic workers' compensation seminars for clinicians and staff, and independent medical examiners training. In addition, the groups offer provider input through systems of sub-committees on governmental affairs, claims issues, care guidelines, and provider education. These committees also serve as an important point of contact for community physicians to obtain clarifications and provide input on department policy as well.

9.4.4 Internal Stakeholders

Similar communication and committee mechanisms are also used to communicate and coordinate activities of various divisions within the agency. A unique feature of the Washington State system is that industrial insurance, research and quality improvement efforts, and workplace health and safety activities exist within the same agency. This provides a ready opportunity to identify policy needs that help formulate research questions relevant to business, labor and agency concerns. For example, research on occupational exposures and primary injury/illness prevention occurs as well as research on medical care, secondary disability prevention and health services research related to workers' compensation. Regular formal and informal interactions among divisions have fostered a constructive environment that has helped to ensure the success of many of the department's projects.

9.5 RESEARCH PROJECTS AND RELATIONSHIPS TO POLICY DEVELOPMENT

Conducting both original research and pilot tests of innovative ideas is crucial to the DLI's approach to policy development. Wide dissemination of findings through regional and national forums provides independent review and comment on this work that offers additional insight and perspective. Staff from several DLI programs, agency actuaries, and the Department of Environmental and Health Sciences at the University of Washington routinely engage in, and collaborate on, projects. Areas of research have included epidemiological studies of work injuries, outcomes studies, field studies of system wide procedures and interventions, and health services research including technology assessments and guidelines development. The information from this work helps to better policy development and ensures more appropriate medical coverage decisions. Table 9.1 identifies examples of recent policy questions that led to the establishment of specific research studies and the value they have in informing subsequent policy development in Washington. Brief overviews of several studies we have done and their relevance to workers' compensation and occupational health generally are described below. Somewhat more detail is offered on some of the larger field studies and current community-wide pilots.

Table 9.1 Examples, policy questions and related research.

Policy question or concern	Related research initiative	Policy development implications
Increasing disability from non-traumatic musculoskeletal disorders	Work-related Musculoskeletal Disorders project	A small percentage of claims use the majority of the systems resources. Need to focus primary prevention efforts in numerous industry activities
Can improvements in claim management processes help reduce work-related disability?	Long-term Disability pilot project	Claim management improvements do not impact disability rates, but do improve customer satisfaction. Lead to Occupational Health Services (OHS) project
Can increasing occupational health expertise in delivery of care to injured workers help reduce work-related disability?	Managed Care pilot project	Disability rates are reduced with more occupationally focused delivery and improves employer satisfaction, but limiting worker choice reduces worker satisfaction. Lead to OHS project
Can increasing occupational health resources and expertise among physicians in the community reduce disability while preserving worker choice of provider?	Occupational Health Services pilot projects	Currently being studied
Does spinal fusion surgery improve outcomes of workers with occupational low back injuries?	Spinal Fusion Outcomes studies	Outcomes of workers undergoing fusion surgery are worse than those who do not. Lead to treatment guidelines outlining patient selection criteria and informed consent processes for workers considering this option
Increases injuries related to the use of pneumatic nail guns in construction industries	Pneumatic Nailer study	Work with industries and manufacturers undertaken to enhance training on safe and proper use and highlight design issues

9.6 EPIDEMIOLOGICAL STUDIES

9.6.1 Factors Influencing Work-related Disability

A population-based retrospective cohort from a random sample of more than 28,000 workers' compensation claims in Washington State was conducted in order to determine if factors predictive of work-related disability could be identified (Cheadle *et al.*, 1994). The principal outcome measure was length of time for which compensation for lost wages was paid, used as a surrogate for duration of temporary total disability. Even after adjusting for severity of injury, the only factors that demonstrated a robust ability to predict longer durations of disability were older age, female gender, and a diagnosis of carpal tunnel syndrome or back/neck sprain. Some other predictors with lower magnitudes of effect included being divorced, working in a firm with less than 50 employees, high county unemployment rates, and work in the construction or agricultural industries. This work helped inform other research, such as the Long-term Disability, and Managed Care Pilot projects described below, that have attempted to target more system-wide interventions to reduce disability.

9.6.2 Work-related Musculoskeletal Disorders of the Neck, Back and Upper Extremities

The Work-related Musculoskeletal Disorders (WRMD) of the Neck, Back and Upper Extremities study examined frequency, incidence rate, severity, cost and industry distribution on non-traumatic soft tissue (NTST) musculoskeletal disorders in the state. The results have been used to help focus prevention efforts both by the DLI and industries in the private sector (Silverstein and Kalat, 2000). State Fund and self-insured workers' compensation claims for non-traumatic (i.e. conditions of gradual onset, rather than linked to a specific injury or event), soft-tissue hand/wrist, elbow, shoulder and back disorders between 1990-1998 were evaluated. These claims made up 26% of State Fund claims and more than 36% were associated with compensable time-loss. The 392,925 State Fund claims accounted for $2.6 billion in direct costs and 20.5 million lost workdays over the period. The overall claims incidence rate was 355 per 10,000 full-time equivalent (FTE) employees and averaged 43,658 claims per year.

The average cost of an NTST claim was $5,923 and compensable claims averaged 146 days of time loss each. Backs accounted for almost 55% of all NTST claims with upper extremity disorders making up 34% of the total. An index was calculated to identify which industries were at highest risk for these problems. Interestingly, the study also identified that temporary help workers were at increased risk compared to industry averages overall. This kind has helped target primary prevention efforts in a number of industries.

9.6.3 Occupational Injuries Among Adolescents

Miller (1995) analyzed workers' compensation claims on reported injuries to minors aged 11-17 in the state during a four-year period from 1988 to 1991. During the period studied, 17,800 such claims were accepted. About 88% of injuries occurred among 16-17 year olds, representing 9 claims per 100 workers annually. On the surface, this rate appears similar to that of the adult workforce. However, given that most adolescents work part-time and for only part of the year, the implications are potentially more significant. When adjusted by the number of hours worked, injury rates to adolescents in the state were actually three times higher than for adults. Almost 6% of the injuries were serious involving fracture, multiple injury, concussion, dislocation, or amputation. Most of the injuries were lacerations, sprains/strains, contusions and burns with the upper extremity being the region most frequently affected. The nature of the claims reflects the fact that most adolescents work in the retail trade, primarily food services. Again, awareness of such findings helps to inform technical assistance activities with employers and can assist in prioritizing primary prevention strategies.

9.6.4 Fatality Assessment and Control Evaluation

Research on claims data identified 335 fatal and 1,105 severe non-fatal injuries in Washington State between 1991 and 1995 (Alexander, *et al.,* 1999). Causes of fatal and non-fatal severe injuries were found to be notably different across industrial risk classes, emphasizing the need to consider more than fatality data. This kind of work has helped establish an ongoing monitoring and intervention program, the Fatality Assessment and Control Evaluation (FACE) Project. FACE collects basic information on all work-related acute trauma and fatalities in the state. Information is compiled from DLI databases as well as the Department of Health, public safety officials, newspapers, medical examiners/coroners reports, and other sources. The information includes worker and employer demographics and preliminary cause as well as a short description of the incident. Research investigators follow-up with more detailed evaluation on a select number of fatal incidents then synthesize the information to develop detailed reports for widespread, but focused, distribution aimed at prevention of future work-related fatalities.

This program developed a comprehensive data collection system and a set of investigative reports that has led to determining root causes of a number of fatalities in the work place. The detail of the researchers' investigations go into to greater depth than those from the health and safety compliance programs, yielding more technical information that can be used by participants to make improvements. As research initiatives that do not have direct enforcement or penalty consequences related to them, participants' anxiety can be reduced. This has permitted the ability to pilot test workplace modifications, such as ergonomic interventions and controls as well as work site education programs. In addition, by pursuing comprehensive assessments and dissemination of findings to various equipment manufacturers, safer redesigns of some equipment may be possible as well.

9.6.5 Pneumatic Nail Gun Injuries

Between 1990 and 1998, there were 3,616 accepted Washington State Fund claims for injuries associated with nail guns, particularly in the construction industries (Baggs *et al.*, 1999). Over the time period, the number of claims more than doubled. Pneumatic nail guns have become increasingly popular in wood frame construction for their contribution to decreasing task times and reducing repetitive stress from hammer use. Fingers and hands made up the bulk of body parts injured and a comprehensive review of how injuries occurred and contributing factors led to the development of a series of specific engineering and administrative controls identified to enhance safe use of pneumatic nailers (Baggs *et al.*, 1999).

9.6.6 Traumatic Head and Brain Injuries

Although head and brain injuries account for only about 100 of the 180,000 workers' compensation claims in the state annually, they are among the most devastating in terms of their human cost, not to mention the burden on the worker's family, employer and social services (Cohen *et al.*, 1999). Examination of Washington State workers' compensation claim data initially identified industries with high incidence of head trauma (Heyer and Franklin, 1994). Subsequently a more comprehensive traumatic head and brain injury (THBI) surveillance project was undertaken to determine if risk factors and strategies for prevention could be identified (Cohen, 1999). Of the 8.6/100,000 FTE overall injury rate, wood frame building construction (specifically metal siding and gutter installation), logging, and grain mills are the industries with the highest incidence rates with 410, 320, and 270/100,000 FTE respectively. Again, this work offers insight for targeting primary injury prevention efforts by employers, labor, and regulators.

9.7 OUTCOMES STUDIES

9.7.1 Lumbar Fusion

A large population-based cohort of injured workers in Washington State who underwent lumbar fusion surgery for low back and leg pain in 1986-87 were evaluated in terms of their work disability status, reoperation rate, and patient satisfaction (Franklin *et al.*, 1994). The purpose of the study was to determine if predictors for outcome from the surgeries could be identified in order to design clinical guidelines. The incidence rate for the procedure was 41.7/100,000 FTE workers annually, translating into 388 patients in the time period studied. 68% of workers receiving fusion surgery remained work-disabled two years after the procedure. Additional surgeries were performed on 23%. Nearly 68% of patients receiving fusion surgery reported that their back pain was worse overall and over 55% indicated that their quality of life was no better or worse. Severity markers including older age at surgery, prior back surgeries, longer time until surgery, longer time-loss durations before surgery, and greater number of levels fused, all predicted worse outcomes from the procedure in this population.

The findings from this work prompted a collaborative effort to work with the Washington State Medical Association's Industrial Insurance Advisory Committee to develop treatment guidelines and informed consent processes to better inform doctors and workers considering lumbar fusion in the absence of spinal fracture. The guidelines describe options for appropriate trials of conservative care and psychological assessment of patients who have been on disability (WSDLI, 1999a,b,c). Additional criteria including condition profile (e.g. instability, radiculopathy) and confounding variables (e.g. drug use, deconditioning, multiple levels) and follow-up are delineated in the guidelines.

The impact of the guidelines and the educational effort regarding the poor outcomes of fusion led to a significant reduction in the lumbar fusion rates in Washington State in the early 1990s (Elam *et al.,* 1997). However, more recent data have indicated that a dramatic rise in fusion rates is occurring, solely related to the introduction of a new fusion device, the interbody cage (Franklin *et al.*, 1998).

9.7.2 Thoracic Outlet Surgery

Thoracic outlet syndrome (TOS) is a condition that has met with controversy and debate within the field of neurology due to lack of diagnostic specificity (Franklin *et al.*, 2000). Thoracic outlet surgery has been increasingly performed in workers' compensation cases where upper extremity symptoms have failed to respond to usual care. A study to evaluate outcomes from this procedure identified 158 Washington State workers' compensation cases between 1986-1991 that underwent thoracic outlet surgery. Disability status was determined from administrative records and telephone interviews were conducted in 1993 with 63% of these individuals to assess current work status, functional outcomes, and satisfaction with the procedure. A sample of 95 workers from 1987-1989 with a TOS diagnosis who did not receive surgery were identified as a comparison group.

On average, individuals first received a diagnosis of TOS 2.4 years after a precipitating injury for which a workers' compensation claim was accepted. The average age at time of injury was 32 years old and 60% percent of patients who underwent the surgery were work-disabled one year after surgery with half of those disabled at least two years. Two-thirds also reported significant, persistent symptoms with 72% indicating they were limited in vigorous activities and 44% were unable to work. The strongest predictors for disability were the amount of disability prior to surgery and a longer time between injury and surgery.

Through collaboration with the DLI's medical advisory committee, work ensued on establishing treatment guidelines which could help predict which workers, under what circumstances, might experience benefit from TOS surgery. Clinical criteria for arterial, venous, and neurologic etiologies of TOS were developed that included specific subjective characteristics, examination findings, and definitive vascular and/or neurodiagnostic studies that were needed prior to authorizing surgery.

9.8 FIELD/SYSTEMS STUDIES

9.8.1 Long-term Disability Pilots Project

The Long-term Disability Prevention Pilot project was an administrative intervention that randomized more than 8,000 employers with 10,000 associated workers' compensation claims in two regions of the state into two different claims administration groups following them over a four year period (McDonald *et al.*, 1998). The two regions of the state that were selected represented different industrial profiles, unemployment rates, mixes of seasonal versus year-round employment, labor markets, and different rates of participation in employer incentive programs. Utilizing the two regions assured representation of urban and rural environment and allowed for more generalizability of the results state-wide.

The "intervention" group consisted of "intensive claim management" involving reduced caseloads and early access to services, while the comparison group consisted of standard claims management in use at the time. The proportion of workers in the intervention group receiving time loss compensation was 22%, which was similar to that in the comparison group (20%).

Disability prevention, return to work and costs were tracked via actuarial data normally maintained in department and state-wide databases. In addition, numbers of protests, disputes, and appeals were tracked. Customer satisfaction surveys were also administered four months after claims were initiated to assess impact on both worker and employer satisfaction with claim management.

Although the more rural and agricultural region demonstrated a small reduction in disability rates in the first year, similar reductions were not seen in the more urban region of the state and were not sustained in subsequent years. A similar result was found with return to work data. The number of workers back at a job increased slightly in the more agricultural region in the first year, but this change was not sustained in subsequent years. No differences in either region were seen in medical or compensation costs.

The intervention group did demonstrate first year improvements in reducing the number of disputes filed by workers by about 25%. In addition, employer protests decreased by 37% in the intervention group. On a four-month, post-claim satisfaction survey, worker satisfaction with claim management was high in 56% of respondents in the control group compared to 76% in the intervention group in more urban region. Worker satisfaction was high in about 74% of respondents in both groups in the more agricultural region of the state. Employer satisfaction with claim management improved in both regions in the intervention groups by 13% and 17% in the agricultural and urban regions respectively.

Although more detailed cost-benefit analysis remains to be done, the studies do indicate that reducing claims loads and engaging in more intensive claim management activity do not impact disability from work injury. Still much has been learned regarding sources of satisfaction with claim processes and this information is being used to enhance existing claims processes and inform strategic planning for future department customer service efforts.

9.8.2 Workers' Compensation Managed Care Pilot Project

In 1993, as part of a state-wide health reform effort, Washington State initiated a pilot project to assess the effects of treating injured workers through managed care approaches. The managed care pilot (MCP) study evaluated the effect of making two substantive changes to the existing fee-for-service delivery system. First, the existing fee-for-service method of payment was changed to an experience rated capitation with participating occupational health plans assuming financial risk for services by agreeing to accept a pre-paid amount for each covered worker (i.e. a set dollar amount per covered-life for all workers, regardless of services rendered to those who become injured). Second, the delivery of care was changed from the existing worker choice model to one in which the workers went to a designated physician network working under the direction of an occupational health physician. The project required a time-limited legislative exemption from existing workers' compensation laws to allow these changes.

Two health plans that offered a high level of occupational health care expertise and organized delivery strategies that could meet the unique needs of injured workers and employers were selected through a competitive bid process. Employers in the region of the selected plans were recruited for voluntary participation. In order for a company to be enrolled as a pilot managed care company, a majority of the companies' employees had to vote in favor of participation. Approximately 7,000 workers at 120 employers participated. Injuries were followed for an 11-month period. Workers in the participating companies agreed to be seen only by the participating occupational health plans for a period of nine months following their injury. After that timeframe, any workers with open claims would revert to the regular fee-for-service mechanism and could go to any provider. For comparison, the 120 managed care firms were matched to 396 comparison firms (12,000 workers) whose injured workers received standard fee-for-service care. Injured workers treated within the occupationally focused managed care plans were compared to fee-for-service injured workers with regard to satisfaction, health outcomes, and medical and disability costs.

There were no meaningful differences in health outcomes between workers enrolled in the occupational health, managed care plans and those under traditional fee for service (Keys *et al.*, 1999). However, important and statistically significant differences favoring managed care patients were found in medical and disability costs (Figure 9.1) (Cheadle *et al.*, 1999). Medical costs per claim were 22% lower, on average, for managed care patients ($587 versus $748). The reduction in disability costs was even greater, primarily due to fewer managed care patients being placed on disability (in Washington, disability payments begin after four days of lost work time). Whereas 14.7% of injured workers under managed care received time loss payments, 19.2% of fee-for-service received them. In addition, those managed care patients who did receive disability incurred lower total disability payment costs ($2,332 versus $3,446) than comparison patients.

It is important to note that the managed care plans were only at risk for medical costs. Disability payments were made by the department in the usual way and were not calculated into the capitated rates. As no financial incentives existed for disability and time loss costs, it seems likely that the occupationally focused nature of the plans would be responsible for the difference rather than any inherent attributes of a managed care delivery model. We believe that the improved integration and coordination of care under the occupational health model along with more frequent employer communication concerning patients' medical status and their need for job modification can facilitate timely return to work (Wickizer *et al.*, 2001). This finding is consistent with other studies that explored occupational health interventions (Loisel *et al.*, 1997).

The cost savings achieved through managed care came at the price of reduced patient satisfaction. Managed care patients were less satisfied than their fee-for-service counterparts with overall care, however, disparity in satisfaction was even greater regarding access to care. Restrictions on choice of provider and the limited number of occupational health care clinics operated by the managed healthcare seem likely to account for these differences.

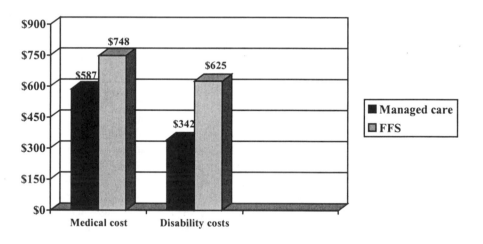

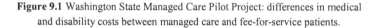

Source: Washington State Department of Labor and Industries

Figure 9.1 Washington State Managed Care Pilot Project: differences in medical and disability costs between managed care and fee-for-service patients.

Early return to work has tangible cost benefits but the question of attendant long-term consequences for health and employment outcomes also needed to be explored. A two-year follow-up study was conducted to assess the long-term outcomes as well as to compare medical and disability costs of workers who had been treated under the two different systems. No evidence was found of any significant long-term differences in either group (Keyes, *et al.*, 2001). At two years post-injury, functional status, level of employment, and wages among managed care and fee-for-service patients were similar. Both groups experienced increases in medical and disability costs per claim from the end of the pilot study to the two-year mark. However, the magnitude of increase was substantially greater for those workers who had been under the fee-for-service model during the study. Post-study medical costs for those patients initially enrolled on the managed care group rose from $587 to $678, a 16% increase. For those who had been in the fee-for-service group, post-study medical costs increased by 25% from $748 to $934 (Wickizer *et al.*, 2001). Disability costs had a similar pattern of increase with post-managed care group disability costs increasing by 21% from $342 to $414. The disability costs among fee-for-service patients rose from $625 to $922 reflecting a 48% increase (Wickizer *et al.*, 2001). A key implication is that patients managed initially under an occupational health model experienced a sustained benefit compared to those who were not.

This pilot study was not conducted in a randomized fashion, hence one cannot discount the possibility that some confounding factors may have affected costs. Because randomization was not feasible, firms from which comparisons were made were carefully matched based on factors known to affect injury risk and costs. In addition, the mix of injuries was similar in both groups (Cheadle *et al.*, 1999). Thus, there is no strong reason to believe that the observed differences resulted from some underlying bias in the analysis.

9.8.3 Occupational Health Services Pilot Projects

Information gained in the Long-term Disability and Managed Care pilot studies presented a significant challenge to the department. The Long-term Disability pilots suggest that administrative interventions and reduced work load on claims staff may help satisfaction levels, but such practices are unlikely to have a sustainable impact on what matters most, worker disability. The managed care data, however, did demonstrate that sustainable reductions in disability could be realized if providers with increased knowledge about occupational issues could navigate the system. Yet managed care arrangements involving restricted physician networks came at the price of reduced levels of satisfaction. Further, there is little doubt that a state that allows workers to choose providers would not be politically acceptable in Washington.

In order to identify approaches and program design options that could preserve choice (a correlate of high patient satisfaction), and increase the occupational health know-how of doctors (a source of high employer satisfaction and disability reduction), a new set of projects have been initiated called "Occupational Health Services (OHS) pilot project". The OHS pilot is a multi-year effort with distinct phases and sub-projects that will develop ways to efficiently track patient outcomes, worker and employer satisfaction, to assess care quality in an ongoing fashion, and to increase knowledge and access to occupational health resources among doctors who treat injured workers (Wickizer *et al.*, 1997; 2001). It is one of the most ambitious studies the department has undertaken and will monitor changes in outcomes and satisfaction within a targeted community. Particular focus is on clinical outcomes for three common occupational conditions: low back pain, carpal tunnel syndrome and extremity fractures. These conditions together represent about 50% of the costs for claims that go on time loss. In addition, the mix of conditions reflect an example of a fairly straightforward acute injury (extremity fracture), a somewhat more complex and disabling condition for which work causation is identifiable (low back problems), and a condition which has a more gradual onset with adjudication challenges related to establishing work-relatedness (carpal tunnel syndrome).

A key aspect of the OHS pilot involves establishment of resource centers in two pilot communities in the state staffed by respected opinion leaders with expertise in occupational health. Continuing education and incentives for taking advantage of the program (both financial and administrative) are being explored. In addition to identifying and developing best-practice strategies to facilitate innovations at a broader community level, methods to track worker and employer satisfaction, functional outcomes, and indicators of quality of care are being developed. Through collaboration and agreement with the Workers' Compensation Advisory Committee, a set of principles, mutually agreed upon by business and labor were as follows:

- Expand capacity for occupational healthcare delivery.
- Increase provider accountability for delivery of efficient and effective care with improved outcomes.
- Improve worker and employer satisfaction.
- Retain the voluntary nature of the workers' current ability to select providers.
- Economic incentive can be used to encourage participation.

9.8.4 National Trends Study

Just prior to the implementation of the formal OHS pilot, a preliminary study was undertaken to gather information on existing innovative occupational health programs around the country (Schulman and Schwartz, 1997). A University of Washington research team conducted key informant telephone interviews with occupational health analysts, program managers, researchers and administrators from around the country. Information was gathered on the structure of the service delivery systems being used, approaches to primary and disability prevention, return to work strategies, approaches to dispute resolution, performance assessment, and what kinds of administrative and financial incentives were in use to encourage patient and employer participation in programs with high levels of occupational health expertise. This study design was qualitative key informant interview, therefore no attempt was made to rate how the various approaches performed compared to each other, but a number of useful insights emerged (Schulman and Schwartz, 1997).

Many informants stressed the importance of early intervention to reduce the risk of long-term disability. Informants also consistently noted the importance of return to work as the key priority in the development of new occupational health delivery programs. They also indicated their recognition of the importance of primary injury prevention, however they noted that key barriers to substantively implementing such programs included high resource costs and inadequate incentives for conducting primary prevention activities. Informants also stressed the value in having physicians with extensive occupational health expertise overseeing the delivery of care within an organized occupational health network. Fashioning appropriate incentives to encourage provider and patient participation in these networks was also highlighted.

9.8.5 Worker Satisfaction Surveys

A worker satisfaction survey was developed and conducted during spring 2000 on approximately 800 injured workers at about four months following their filing of a workers' compensation claim. Using structured telephone interviews, workers' satisfaction with various dimensions around the episode of care they recently had for their injury was determined. The survey showed that workers in general were quite satisfied with different aspects of the episode of care. For example, 83% of the workers surveyed reported having very little difficulty in getting initial medical care needed to treat their injury. Overall 73% of the workers were very satisfied with the quality of care they received, and 87% indicated they felt their healthcare provider was courteous and treated them with respect. A smaller percentage (59%) was very satisfied with how their care was coordinated.

The survey identified other areas of care delivery with which workers were less satisfied, such as 27% of survey respondents referred to specialists indicated their referral was not made soon enough. Only 20% of the workers reported that their doctor or healthcare provider talked with their employer about return to work issues. The results of this worker satisfaction survey will serve as a baseline for future assessments regarding the impact that any delivery system changes in the pilot communities have, as well as inform planners about where quality improvement efforts may be needed. This recent satisfaction survey generated different results from the earlier managed care patient satisfaction survey. Even among the fee-for-service patients who served as controls for the managed care pilot, satisfaction with access to care and access to attending physicians was 10% to 20% points lower. The reason(s) for the differences is unclear. Methodological differences in the administration and construction of the surveys as well as wording of questions could account for some of the difference. Also, differences in the underlying populations could have accounted for some of the differences in results.

9.8.6 Employer Satisfaction Surveys

An employer satisfaction survey has also been developed as part of the OHS project and focuses on physician-employer communication. In the managed care pilot, the frequency and quality of communication occupational health physicians had with employers was a source of high employer satisfaction. In the OHS project, care coordination, return-to-work focus and communication with employers represent key delivery system improvements that will be implemented. This instrument expands on areas of concern to employers and will help assess how employers in the pilot regions react to changes in the delivery system. It is expected that more frequent employer-provider communication and interaction will result in improved disability prevention and more timely return to work.

9.8.7 Outcome Tracking

Self-reported functional status using questions from validated instruments for the three conditions under study (e.g. SF-36, Roland-Morris back questionnaire) are being used to assess outcomes of care. Key data elements related to return-to-work will also be gathered. Outcomes tracking will be prospective in design with claimants being initially contacted by telephone within the first few weeks following injury and being re-surveyed six months later. In addition to providing important outcome benchmark information for the quality improvement initiative, the outcome survey will enable validation of another key dimension of the project, quality of care indicators.

The outcomes tracking survey will assist in accomplishing an additional important project goal. Relying solely on outcome surveys for routine evaluation and patient monitoring is impractical due to budget and time constraints. Efficient, low cost methods have to be identified that can be used on a wider, ongoing basis to assess patient outcomes, e.g., disability days and return to work. Administrative data that are already routinely collected in the course of workers' compensation claims represent a source of readily available information. The DLI maintains data on time loss days for all workers' compensation claimants and another agency, the Washington State Department of Employment Security, maintains data on employment and earnings. Correlations will be performed between the self-reported outcomes data and these administrative databases. In this way, the feasibility of using a mix of administrative and self-report data for monitoring patient outcomes can be determined.

9.8.8 Quality Indicators

Quality of care is another dimension of delivery that is being explored. Quality indicators for each of the three conditions being followed (low back sprain, carpal tunnel and fractures) are under development. The research team developed a seed list of possible quality indicators based on reviews of clinical, quality improvement, and scientific literature. The indicators, along with supporting research, are given to a team of community physicians and nationally recognized experts in the field. Through a nominal group process, the indicators are rated according to their perceived utility for quality improvement by the clinicians, and as possible performance measures.

Several quality indicators were developed this way for occupational carpal tunnel syndrome and low back pain:

Carpal Tunnel Syndrome

- The exam (history and physical findings) screens for the presence/absence of Carpal Tunnel Syndrome (CTS) on the first healthcare provider visit for suspected occupational CTS.
- Specialist referral is made if no improvement in symptoms, continued time loss, or diagnostic uncertainty exists by six weeks after first visit.
- Assessment of (or referral for assessment of) impediments to return to work by the fourth week of work loss.
- Diagnostic plan documented at first or second health care provider visit.

Low Back Pain

- The exam (history and physical findings) screens for the presence/absence of signs of radiculopathy and other critical neurological findings at the first healthcare provider visit for suspected occupational low back pain (LBP).
- Treatment plan is documented at each provider visit.
- Activity prescription for patient is given, discussed, and documented at each provider visit.
- If not returned to work or part-time work, patient has healthcare provider visit scheduled at least every two weeks with at least one visit scheduled two to four weeks after return to work.

The indicators will be useful in several ways including offering feedback to individual clinicians and the community resource centers for shaping continuing education and mentorship training. In addition, certain measures such as documentation of timely clinical screening procedures in patient charts may serve as performance benchmarks in center contracts and/or provider agreements.

Community-based Resource Centers of Occupational Health and Education

Implementing effective delivery system improvements will require partnerships and coordination between providers, labor, employers, and the workers' compensation payers and regulators. Delivery system improvements will focus in part on increasing the occupational health expertise of attending physicians who only occasionally see injured workers, improving coordination of care and return to work strategies, and establishing administrative and financial incentives that assure appropriate services are reimbursable.

Care delivery processes for injured workers are currently underdeveloped. Practices among physicians exhibit large variations due to a lack of interest in and knowledge about occupational health best-practices. Most workers' compensation fee schedules and payment policies do not offer incentives for physicians to improve their skills in treating workers with occupational conditions. By developing formal agreements and appropriate positive incentives with providers, the DLI plans to encourage physicians to gain skills such as developing a return to work plan with the worker and employer, referring quickly to clinical specialists when progress becomes stalled or risks of disability are identified, and regularly communicating with the employer to facilitate return to work.

9.9 HEALTH SERVICES RESEARCH

9.9.1 Resource Costs of Chiropractic Services

With the adoption of a resource-based, relative value system for reimbursement being adopted by the US Health Care Financing Administration in 1992, a new era of "evidence-based" reimbursement began in American healthcare payment policies. Reimbursement scales are now set by assessment of what the resource costs (physician work, practice overhead, and malpractice risk) are for every procedure. Chiropractic services were not included in the work done initially at the federal level. The DLI payment policy adopted Relative Value System methodology and given that more than half of occupational low back patients seek these services, a project in collaboration with health services researchers at the University of Washington, the DLI, and representatives from the Washington State Chiropractic Association was established. The project obtained private foundation support to conduct a national randomized survey of chiropractors using HCFA's methodology to assess the work done by chiropractic physicians (Hess *et al.*, 1997; Hess and Mootz, 1999).

Based on findings from this work, it was determined that DCs, DOs, and MDs all perform similar amounts of work in similar patient work-ups. A new fee schedule for chiropractic services was developed that included reimbursement for services previously not allowed and implemented a new set of codes that accurately characterize and reflect the work performed in follow-up care encounters.

9.9.2 Reliability of Pulmonary Impairment Rating

Impairment rating regulations for the DLI were established in 1974, including some specific to Washington State for respiratory and spine conditions. As part of overall regulatory and quality improvement efforts, and in response to many requests, the rating system used for pulmonary permanent disability was evaluated for its reliability. Inter- and intra-rater reliability tests were performed and the established system was found to be problematic. As a result, a new system was developed based on methods that had better reliability and new rules were established in 1991 incorporating the improvements.

9.9.3 Technology Assessment of New Medical Devices

In industrial insurance, medical care may be constrained only by the achievement of some level of functional status of the patient (e.g. return to work, pre-injury status, maximal improvement) rather than a limit on expenditures or services. A fundamental challenge for policy makers is to assess relative merits of various procedures, devices and drugs. Unlike bringing new drugs to market, new medical devices can rapidly escalate into use without clear evidence of their efficacy for given clinical indications. The United States Food and Drug Administration (FDA) often extends approval for marketing new medical devices based on "510(k) equivalency" which only requires that a proponent demonstrate that the device is fundamentally similar to another approved product in terms of its technical and safety characteristics. There is virtually no FDA requirement that such new medical devices demonstrate any kind of clinical efficacy in order to be marketed to healthcare providers for a given clinical application (Ramsey *et al.*, 1998).

As a result, development and implementation of a formal technology assessment process was undertaken by the DLI to better incorporate scientific evidence, expert opinion, and community opinion into global coverage decisions made by the department for medical devices (Franklin *et al.*, 1998). The process for evaluation of a medical device may be initiated through application by any interested party (provider, equipment manufacturer, claim manager, worker or employer) by submission of a Medical Device Coverage Request (http://www.lni.wa.gov/omd/healthpol/techas.asp). Obviously the number of devices that could need formal evaluation might readily exceed department resources available for such assessments. Thus, evaluation of devices by the DLI is prioritized based on perceived clinical need, availability of alternatives, community demand, published scientific literature, costs, and other factors.

The DLI has encountered numerous cases of exaggerated claims for effectiveness, or use of a new device as a substitute for an existing standard procedure (Franklin *et al.*, 1998). For example, screening quantitative (psychophysical) sensory tests, such as neurometry, have been billed in place of standard nerve conduction velocity tests and electromyographic scanning using surface electrodes has been billed under approved needled electromyographic services. Neither of the newer technologies has the localizing value or specificity of the existing tests and could not be recommended as replacements (Fanzblau *et al.*, 1994, Haig *et al.*, 1996, Gelber *et al.*, 1995). The technology assessment process has helped inform the overall process for making appropriate medical device coverage decisions. Technical assistance for technology assessment processes has also been done including training, through collaboration with other state health insurers, and publications in clinical literature (Hansen and Mootz, 1996). As better understanding of what constitutes utility, quality, efficacy and value for patients translates into medical coverage policy, an enhance ability to fulfill the DLI's statutory mandate to assure quality care for workers can be enhanced.

9.10 SUMMARY AND CONCLUSION

Research and policy activity in Washington State has been interrelated in substantive ways in recent years. Policy needs such as those arising from poor quality of care, high injury rates, or excessive disability rates have driven innovative research initiatives. Collaborations with researchers at the University of

Washington, representatives of the business and labor communities, providers, and various private sector organizations have lead to practical projects that yield information that can be used to improve the workers' compensation system.

The identification of key problems that result in poor care quality and/or excessive disability among injured workers, and implementing system and delivery changes that may address these problems and improve quality of occupational healthcare, is an ongoing task. Making such improvements is tedious and fraught with complexities that require balancing interests of scientific integrity, political realities, resource constraints, and the institutional inertias that slow change in the first place. In an environment of increased public scrutiny of payers and providers, evidence-based decision-making for health policy is becoming mainstream.

The approach being taken in Washington has focused on the identification of policy needs within the system, putting research teams and resources together to design ways to assess the underlying attributes of the problems, and pilot test innovative strategies to resolve them. The yield in new information, and the constructive nature of collaboration with the system's customers and players has allowed improvements to occur. Innovation and systematic testing of new ideas to decide what to bring online or where to go next is a hallmark of successful organizations. Although our efforts occurred in the absence of disagreement or controversy, our experience underscores the importance of establishing broad-based support particularly for delivery system innovations.

Our research activities also benefited from the close collaboration between program staff and health services researchers at the DLI and the University of Washington. The DLI staff bring important program and policy experience, along with an appreciation of the context and environment within which the research, policy and research and development activities need to be conducted. Researchers at the DLI and the university bring scientific rigor and methodological expertise to the project design, implementation and interpretation.

The DLI represents a "single payer" for purposes of workers' compensation. Washington State, along with five other states, has a state-fund system that requires all employers that are not self-insured to purchase workers' compensation insurance through the State Fund. No matter what one feels about the merits or drawbacks of a single payer system for purposes of financing healthcare, the fact is that such a system creates important opportunities for policy initiatives and for research and evaluation. Nonetheless, experience and findings from our epidemiological studies, work place safety and exposure efforts, outcomes studies, field-based pilots, and health service research work have generic applicability.

General medical care shares many of the same problems and challenges as occupational healthcare with regard to quality and care coordination. Care for patients with chronic diseases, such as diabetes, is often fragmented. Issues of provider accountability, misdirected incentives, and ineffective clinical management processes erode quality and compromise health outcomes for general medical care patients as well as ill or injured workers. These kinds of innovative quality improvement efforts have been implemented in some patient populations within closed systems, such as the program to improve care for diabetics at the Group Health Cooperative in Seattle, a large staff model HMO (McCulloch *et al.*,

1998). However, such efforts have rarely been attempted on a community-wide basis. Our current OHS pilot project represents an ambitious community and systems "intervention" effort that has evolved from a decade of smaller research and policy initiatives, each pointing to a better solution. This project, and the approach in general, provides an example for how valuable experience can be gained through research for developing strategies to improve quality and health outcomes on a community basis.

9.11 KEY MESSAGES

- Washington State provides a unique model of close collaboration among governmental regulatory authorities in occupational health and safety, a single public-sector workers' compensation insurer, private and public sector employers, unions and employee associations, and various elements of local healthcare system, linked closely to university-based researchers.

- The result has been a very productive program of large-scale observational and intervention studies documenting "what works" for reducing work-related disability in this setting, as well as what outcomes appear resistant to impact by new disability management and care-organizational strategies.

- Novel outcome studies for expensive and aggressive surgical procedures in injured workers (spinal fusion, thoracic outlet surgery) have led to practical clinical and reimbursement guidelines to improve case selection for surgery and thus improve appropriateness, as well as quality of care.

- A large community trial of intensive claims management for disability reduction did not reduce lost time *per se*, but it did increase all stakeholders' satisfaction with the compensation process.

- A second community trial of "managed-care" by capitation-reimbursed preferred providers with special occupational health expertise, rather than usual fee-for-service care, showed no impact on disability duration or final health outcomes for injured workers. However, medical costs were reduced 22% and disability costs fell even more, due to reduced lost time on the job. Notably, these gains came at the cost of significantly reduced claimant satisfaction under managed care.

- A new quasi-experimental program of studies now under way will utilize multiple strategies to identify and replicate "best practices" for disability reduction in two pilot communities.

9.12 REFERENCES

Alexander, B.H., Franklin, G.M. and Fulton, K.D., 1999, Comparison of fatal and severe nonfatal traumatic work-related injuries in Washington State. *American Journal of Industrial Insurance*, **36**, pp. 317-325.

Baggs, J., Cohen, M., Kalat, J. and Silverstein, B., 1999, *Pneumatic Nailer ("Nail Gun") Injuries in Washington State 1990-1998. Techical Report No. 59-1-1999.* Olympia, WA: Safety and Health Assessment and Research for Prevention Program, State of Washington Department of Labor and Industries.

Beals, R.K., 1984, Compensation and recovery from injury. *Western Journal of Medicine*, **140**, pp. 233-237.

Cheadle, A., Franklin, G., Wolfhagen, C., Savarino, J., Liu P.Y., Salley, C. and Weaver, M., 1994, Factors influencing the duration of work-related disability: a population-based study of Washington State workers' compensation. *American Journal of Public Health*, **84**, 190-196.

Cheadle, A., Wickizer, T.M., Franklin, G., Cain, K., Joesch, J., Kyes, K., Madden, C., Murphy, L., Plaeger-Brockway, R. and Weaver, M., 1999, Evaluation of the Washington State Workers' Compensation Managed Care Pilot Project II: Medical and Disability Costs. *Medical Care*, **37**, pp. 982-993.

Cheal, D.D., 1986, *History and development of Worker's Compensation laws in Washington.* Olympia, WA: Washington State Legislature Joint Select Committee on Workers.

Cohen, M., Kalat, J., and Silverstein, B., 1999, Work-related traumatic head and brain injuries in Washington State, 1990-1997. *Technical Report Number 57-1-1999.* Olympia , WA: Safety and Health Assessment and Research for Prevention Program, State of Washington Department of Labor and Industries.

Collins, P.C. and Porras, J.L., 1997, *Built to Last: Successful Visionary Companies.* (New York: Harper Collins).

Elam, K., Taylor, V., Ciol, A.M., Franklin, G.M. and Deyo, R., 1997, Impact of a workers' compensation practice guideline on lumbar spine fusion in Washington State. *Medical Care*, **35**, pp. 417-424.

Fanzblau, A., Werner, R.A., Johnston, E. and Torrey, S., 1994, Evaluation of current perception thresholds testing as a screening procedure for carpal tunnel syndrome among industrial workers. *Journal of Occupational Medicine*, **36**, pp. 879-885.

Franklin, G.M., Lifka, J. and Milstein, J., 1998, Device evaluation and coverage policy in workers' compensation: Examples from Washington State. *American Journal of Managed Care*, 4(SI), SP178-SP186.

Franklin, G.M., Haug, J., Heyer, N.J, McKeefrey, S.P. and Picciano, J.F., 1994, Outcome of lumbar fusion in Washington State workers' compensation. *Spine*, **19**, pp. 1897-1904.

Franklin, G.M, Fulton-Kehoe, D., Bradely, C. and Smith-Weller, T., 2000, Outcome of surgery for thoracic outlet syndrome in Washington State workers' compensation. *Neurology*, **54**, pp. 1252-1257.

Fordyce, W.E., 1995, *Back Pain in the Workplace: Management of Disability in Nonspecific Conditions.* (Seattle: IASP Press).

Gelber, D.A., Pfiefer, M.A., Broadstone, V., Munster, E.W., Peterson, M., Arezzo, J.C., Shamoon, H., Zeidler, A., Clements, R., Green, D.A., Porte, D., Laudadio,

C., and Bril, V., 1995, Components of variance for vibrometry and thermal threshold in normal and diabetic subjects. *Journal of Diabetic Complications*, **9**, pp.170-176.

Haig, A.J., Gelblum, J.B., Rechtiem, J.J. and Gitter, A.J., 1996, Technology assessment: The use of surface EMG in the diagnosis and treatment of nerve and muscle disorders. *Muscle Nerve*, **19**, pp. 392-395.

Hansen, D.T. and Mootz, R.D., 1996, Formal processes in health care technology assessment. A primer for the chiropractic profession. *Top Clinical Chiropratrics*, **3**, pp. 71-83.

Hess, J.A. and Mootz, R.D., 1999, Comparisons of work estimates by chiropractic physicians with those of medical and osteopathic providers. *Journal Manipulative and Physiological Therapeutics*, **22**, pp. 280-291.

Hess, J.A., Mootz, R.D., Madden, C.W. and Perrin, E.B., 1997, Establishment of total and intraservice work by chiropractic physicians for Evaluation/ Management services and Spinal Manipulative Therapy. *Journal of Manipulative and Physiological Therapeutics*, **2**, pp. 13-23.

Heyer, N.J. and Franklin, G.M., 1994, Work-related traumatic brain injury in Washington State, 1988 through 1990. *American Journal of Public Health*, **84**, pp. 1106-1109.

Kyes, K.B. *et al.*, 1999, Evaluation of the Washington State Workers' Compensation Managed Care Pilot Project I: Medical Outcomes and Patient Satisfaction. *Medical Care*, **37**, pp. 972-981.

Kyes, K.B., Wickizer, T.M. and Franklin, G.M., 2001, Two-year follow-up of workers treated in the Washington state workers' compensation managed care pilot project. *American Journal of Industrial Medicine*, **40** (6), pp. 619-626.

Larson, A., 1952, The nature and origins of Workmen's Compensation. *Cornell Law Quarterly*, **37**, p. 206.

Larson, A., 1992, *Workmen's Compensation for Occupational Injuries and Death.* (New York: Matthew Bender).

Loisel, P., Abenheim, L., Durand, P., *et al.*, 1997, A population-based, randomized clinical trial on back pain management. *Spine,* **22**, pp. 2911-2918.

McCulloch, D.T., Price, M.J., Hindmars H.M. and Wagner, E.H., 1998, A population-based approach to diabetes management in a primary care setting: early results and lessons learned. *Effective Clinical Management*, **1**, pp.12-21.

McDonald, C., Wolfhagen, C. and Franklin, G., 1998, *Long-Term Disability Prevention Pilots: 1998 Report to the Legislature.* Olympia, WA: State of Washington Department of Labor and Industries.

Miller, M., 1995, *Occupational injuries among adolescents in Washington State, 1988-1991: A review of workers' compensation data. Technical Report Number 35-1-1995.* Olympia, WA: Safety and Health Assessment and Research for Prevention Program, State of Washington Department of Labor and Industries.

Mootz, R.D., Franklin, G.M. and Stoner, W.H., 1999, Strategies for preventing chronic disability in injured workers. *Top Clinical Chiropractrics*, **6**, pp. 13-23.

Pound, W.T, 1994, *State of Workers' Compensation.* Denver, CO: National Conference of State Legislatures Task Force on Workers' compensation.

Ramsey, S.D., Luce, B.R, Deyo, R. and Franklin, G., 1998, The limited state of technology assessment for medical devices: Facing the issues. *American Journal of Managed Care*, **4**, SP188-SP199.

Rothstein, M.A., 1990. Occupational Safety and Health Law, 3rd ed., St Paul, MN: West Publishing.

Skolick, A.M., 1962, New benchmarks in Workmen's Compensation. *Social Security Bulletin*, **6**, pp 3-19.

Schulman, B. and Schwartz, S., 1997, *Workers' Compensation/Occupational Health National Trends Study*. Olympia, WA: Washington State Department of Labor and Industries.

Sliverstein, B. and Kalat, J., 2000, *Work related disorders of the back and upper extremity in Washington State, 1990-1998. Technical Report Number 40-4a-2000*. Olympia , WA: Safety and Health Assessment and Research for Prevention Program, State of Washington Department of Labor and Industries.

Washington State Department of Labor and Industries, Office of the Medical Director, 1999a, *Medical Treatment Guidelines*. Olympia, WA: Washington State Department of Labor and Industries.

Washington State Department of Labor and Industries, Office of the Medical Director, 1999b, *Attending Doctors Handbook for Doctors and Office Staff*. Olympia, WA: Washington State Department of Labor and Industries.

Washington State Department of Labor and Industries, Office of the Medical Director, 1999c, *Chiropractic Physician's Guide: Workers' Compensation in Washington*. Olympia, WA: Washington State Department of Labor and Industries.

Wickizer, T., Franklin, G.M., Plaeger-Brockway, R. and Mootz, R.D., 2001, Realigning incentives and clinical management processes to improve quality: The Washington State Occupational Health Services Project. *Milbank Quarterly*, **79**, pp 5-33.

Wickizer, T.M., *et al.*, 1997, *Workers' Compensation Managed Care Pilot Project: Final Report to the Legislature*. Olympia, WA: Washington State Department of Labor and Industries.

CHAPTER 10

Economic Incentives and Workplace Safety

Terry Thomason

The problem of work injuries is a substantial one. Recent estimates put the cost of workers' compensation benefits paid to injured Canadian workers and their families at more than $6 billion annually, or nearly 1% of gross domestic product. In the United States, workers' compensation benefit payments amount to over $40 billion annually. And workers' compensation benefit payments represent only a small portion of the economic costs of work injuries. Work injuries also entail losses due to lost production, damage to plant and equipment, and the uncompensated losses suffered by injured workers that are estimated to be as much as four times the cost of benefits (Heinrich *et al.*, 1980).

The remainder of this chapter proceeds as follows. The next section discusses the economic theory of work injuries and illnesses. Specifically, this section examines employer and worker incentives for safety in the absence of government regulation. The chapter then discusses safety incentives created by different types of government regulation. Conclusions are drawn in the final section.

10.1 ECONOMIC THEORY OF WORK INJURIES AND ILLNESSES

Work injuries are an unwelcome by-product of economic activity. In part, they are random events, but they are also, to some extent, under the control of workers and employers. Employers make can reduce the number of workplace injuries and illnesses by investing in safer technology, providing workers with personal protective equipment (such as hard hats and safety glasses), training workers and their supervisors, etc.; workers can avoid accidents by following safe work practices and by taking greater care on the job.

Both parties incur costs when an accident occurs. Workers' costs include potential loss of income and medical expenses associated with treatment and rehabilitation as well as intangibles, such as pain and suffering and disability that reduces the ability to enjoy leisure activities. Employers' costs include interruptions in production and damage to capital equipment and physical plant.

Since accident prevention also entails costs to employers and employees, public policy should encourage employers and employees to minimize the combined costs of accidents and accident prevention that are incurred by both workers and employers.[1] It is possible to spend both too much and too little on accident prevention. Investment in accident prevention is *socially efficient* when total costs are minimized, that is, when an additional dollar spent on prevention reduces accident costs by exactly one dollar.

As indicated, both employers and workers affect workplace health and safety. We can expect that – if they are rational – both actors will make accident prevention decisions that are *privately efficient*. That is, we may expect that each will make decisions that minimize their own accident and accident costs individually; however, their decision making process may not consider costs that are incurred by the other party.

However, under some conditions, it is at least arguable that employers do consider the workers' accident costs when making investments in workplace health and safety and thus make socially efficient decisions as well. To understand this argument, let us consider a world where there are two types of employers, those with safe workplaces and those with hazardous ones. Assume that workers employed by safe firms do not risk having an accident or illness while at work – i.e. the probability of injury or illness is zero – while one of every ten workers employed by hazardous firms will have an occupational accident each year. Let us further assume that workers are aware of the probability of accidents at both types of firms and that they are free to choose the type of firm for whom they will work.

Under these assumptions, we may expect that if everything else were equal – i.e. the compensation package and other terms and conditions of employment – all workers would prefer employment at the safe firms. In order to attract workers, hazardous firms will be forced to increase wages above the level paid by safe firms. In other words, we would expect to find that hazardous firms pay a compensating differential and that the magnitude of this differential will be related to the workers' expected accident costs, including the cost of lost income, medical expenses, pain and suffering, etc.

So, for example, let us assume that the average cost of accidents for workers is $10,000 and that the average annual salary of workers in safe firms is $40,000.[2] Since the probability of an accident at a hazardous workplace is 0.1, then expected accident costs at that workplace are $1,000 (= 0.1 x $10,000). This means that hazardous employers must pay their employees an annual salary of $41,000 for employment at a hazardous firm to be equally attractive as employment at a safe firm.[3] Thus, the employer's accident costs include the expected accident costs borne by workers. Importantly, employers will be able to reduce the compensating differential and, consequently, their accident costs, by reducing the incidence of workplace accidents and illnesses.[4]

The economic model presented in the preceding paragraphs rests on a number of key assumptions, which many have questioned. In particular, the model requires that workers have complete and accurate information with respect to the risk of injury or death and an absence of barriers to worker mobility, i.e. that workers are free to move in and out of the labor market or between employers at relatively low cost. However, critics point out that it is likely that either workers do not have access to good information about injury risks or barriers to mobility prevent workers from moving to safer jobs. As a result, wage differentials due to the risk of injury either do not arise or they are inadequate, i.e. they do not fully compensate workers for the risk of injury.

Do employers, in fact, pay a compensating differential to workers exposed to greater risks of injury or illness? To answer this question the researcher must address a number of methodological issues that are not easy or simple to resolve, and existing statistical evidence is decidedly mixed. By and large, research investigating the relationship between the risk of fatal injury and wages has found a risk premium, while studies examining the relationship between wages and non-fatal risks have not (see Viscusi, 1993), for a recent review of this literature). However, Dorman and Hagstrom (1998) demonstrate that even fatal-risk differentials are extremely sensitive to the regression specification.

Importantly, there is no evidence to demonstrate that the risk differential is fully compensating, even for fatal injuries. In addition, research suggests that, after controlling for the risk of injury and a variety of other factors affecting wages, the wage differential is substantially larger for unionized workers than for non-union workers (Olson, 1981; Dickens, 1984; Fairris, 1992; Siebert and Wei, 1994; and Sandy and Elliott, 1996). This result, which indicates that union workers get a greater premium for the same level of risk, is difficult to reconcile with the hypothesis that wage differentials compensate workers for the expected cost of accidents.[5] Finally, psychological research suggests that people overestimate the likelihood of a low probability event and underestimate the likelihood of a high probability event (Viscusi, 1993). This systematic bias implies that workers will generally demand a risk premium that is less than fully compensating.

10.1.1 Workers' Compensation

Workers' compensation provides cash benefits to workers who are unable to work as the result of an occupational injury or illness as well as medical benefits and rehabilitative services to all who are injured as the result of a workplace accident.

These benefits have the effect of reducing accident costs for workers and, consequently, the risk premium paid by hazardous employers. As a result, we may expect that the worker's incentive for avoiding workplace injuries will have been reduced because their accident costs have been reduced by the medical and cash benefits provided by the workers' compensation program, a problem known as risk-bearing moral hazard in the insurance literature. We might also expect that workers' compensation benefits would increase the workers' willingness to expose themselves to greater risks on the job, but that these benefits would also increase the likelihood that workers would report an injury that would have otherwise gone unreported or even falsely report a non-work-related injury as occupational. This latter problem is known as reporting moral hazard. In either event, because workers' compensation reduces the cost of workplace accidents for workers, we would expect it would also reduce the compensating wage differential. In fact, there is some statistical evidence indicating that as compensation becomes more generous, the risk premium for hazardous work is reduced.

Do workers' compensation benefits affect employers' incentives to prevent workplace accidents? The answer depends on the way in which compensation benefits are funded. If the employer is liable for workers' compensation benefits paid to his or her firm's injured workers, then the employers' incentive structure will be unchanged by the introduction of workers' compensation. However, if there is no relationship between employer costs and worker benefits, then the employer's incentive to prevent accidents is reduced by workers' compensation benefits.

In Canada and the United States, workers' compensation benefits are funded through a payroll tax paid by employers. A two-step process determines tax (or assessment) rates in most provinces. In the first step, industrial classifications are used to group firms who share similar risks of workplace injury or illness, so that banks are grouped with other financial institutions, for example, food stores are grouped with similar retail establishments, etc. The recent historical accident record of each of these classifications, known as rate groups, is used to determine the base assessment rate for each group. The assessment rate is set so as to provide sufficient income to fund all workers' compensation benefits paid to workers and any expenses associated with workers' compensation program administration.

In the second step of the rate-making process, known as experience rating, the base assessment rate for some firms is adjusted to account for the firm's individual safety record.[6] In other words, the assessment rate for firms with better than average safety records (lower injury rates) is reduced, and the rates of firms with worse than average safety records (higher injury rates) are increased.

Both steps of the rate-making process should reduce the injury rate relative to a regime where all employers are charged an identical assessment rate unrelated to the risk of injury. Variation in the base assessment rate means firms in hazardous industries pay a higher base assessment rate than firms in relatively safe ones, so that the cost of goods and services produced by firms in hazardous industries increases relative to a regime in which a flat assessment rate is charged to all employers. In turn, this reduces consumption of goods and services in hazardous industries relative to safe ones and subsequently employment; as the proportion of employment in safe industries rises, the overall accident rate will drop.

However, the base assessment rate is only marginally related to the firm's accident experience. If the firm is not experience-rated, the employer does not consider workers' compensation assessments to be part of the cost of accidents, since it cannot affect costs by preventing accidents. However, if the firm is experience-rated, then a reduction in the accident rate directly reduces its subsequent accident costs. Thus, if the firm is experience-rated, the employers' investment in workplace safety will remain unchanged following the introduction of workers' compensation insurance; however, if the firm is not experience-rated, the employer's safety investment will decline after workers' compensation is introduced.

Thus, workers' compensation unambiguously reduces workers' safety incentives and increases workers' incentives to report compensable claims. Furthermore, since not all employers are experience-rated, the overall impact of workers' compensation is to also reduce, on average, health and safety investments by employers.

10.1.2 Workers' Compensation and Occupational Injuries: The Evidence

Since the introduction of workers' compensation pre-dates the collection of injury rate data, there are only a handful of studies that have attempted to directly examine this issue and those that do have produced contradictory results. Chelius (1976) found that the introduction of workers' compensation programs led to a reduction in fatal accident rates relative to the tort regime that preceded them. However, Fishback (1987) reached the opposite conclusion, using a different (and arguably better) data set.

On the other hand, several studies have attempted to determine whether there is a relationship between the generosity of workers' compensation benefits and the work injury rate. As indicated, economic theory suggests that, where workers' compensation insurance is less than perfectly experience-rated, the accident rate should be positively related to workers compensation benefit generosity.

A large number of studies using different methodologies and data sources have found the expected positive relationship between benefit levels and injury (or workers' compensation claim) rates. Studies of the US workers' compensation include Butler and Worrall (1983) and Chelius (1982) who examined state-level claims and injury data and Hirsch *et al.* (1997) who used longitudinal survey data to estimate the impact of benefit generosity on the probability that a worker would file a workers' compensation claim. Canadian studies include Thomason and Hyatt (1997), who examined provincial injury rates and Thomason and Pozzebon (1995), who used data on individual workers to estimate claim probability. Uniformly, these studies have found that higher levels of workers' compensation benefits are associated with higher injury or claims rates or a higher probability that a worker would initiate a compensation claim.

10.2 GOVERNMENT REGULATION AND WORKPLACE SAFETY

There are at least three approaches to the regulation of occupational health and safety, all of which have been adopted by policy makers in one form or another at one time or another. The first – and the one most commonly identified as occupational health and safety regulation – involves the promulgation of rules prescribing or proscribing specific policies and practices by employers, which are enforced through onsite inspections and monetary penalties for infractions. The second approach comprehends systems of general safety incentives that reward or punish employers on the basis of safety and health outcomes rather than behaviors that are thought to affect those outcomes. This second approach is embodied in the experience rating of workers' compensation assessments, whereby employers' compensation costs are tied to their accident experience. The third approach, termed internal responsibility, pervasive in Canada, is designed to improve safety and health conditions through workers empowerment and involves three principal elements: 1) the worker's right to refuse to perform unsafe work; 2) the worker's right to information on the nature of workplace hazards; and 3) joint labour-management safety and health committees, which are given a mandate to oversee safety and health conditions in the workplace.

10.2.1 The Economics of Regulation

Occupational safety and health regulation seeks to change behavior of the employer by changing the cost-benefit calculus described in the previous section, through imposition of monetary penalties or other sanctions.[7] Specifically, regulatory sanctions lower accident prevention costs by the expected value of the sanction.[8] In other words, firms considering adoption of a particular safety practice must now weigh expected costs of the sanction that will be imposed if they fail to do so. Sanction costs are characterized as "expected" because, under some – if not all – regulatory regimes, penalties are not imposed unless a violation is detected.

There are two costs that must be considered by efficient regulators: the administrative costs of regulation (the cost of staff involved in enforcement and adjudication), and the cost of regulatory error (the imposition of sanctions whose expected costs are either too great or too small). Sanctions are too small (large) if the costs of accident prevention, including the expected savings from the avoidance of sanctions, are less (greater) than associated accident costs. The cost of error is equal to the difference between accident costs and the cost of accident prevention if the regulation in question is adopted and enforced. The goal of efficient regulation is to minimize the sum of these costs.

10.2.2 Direct Regulation of Workplace Hazards

As indicated, direct regulation attempts to change employer behavior by promulgating regulations that prescribe or prohibit specific employer or worker practices. Regulations are enforced through workplace inspections and penalties for non-compliance. Critics argue that direct regulation fails to recognize important variation across firms with respect to technology and other characteristics. In other words, a safety practice that is efficient for one employer may not be efficient for another, so that there are potentially substantial error costs. In addition, a system of direct regulation in which the regulatory agency responsible for promulgating rules is one-step removed from the workplace and is, therefore, slow to respond to technological change. Once again, this could result in substantial error costs.

Furthermore, as Dorman (1996: p. 197) notes, "Most occupational risks are transitory…Safety features mandated by law may be unavailable or malfunctioning from time to time, but inspectors are not likely to know this." In other words, the probability of detecting non-compliance is low so that the regulators must substantially increase the magnitude of the sanction imposed. Finally, direct regulation is costly to administer. It requires an extensive bureaucracy to develop standards, inspect workplaces, and to resolve disputes with employers concerning the appropriateness of penalties.

Most research examining direct regulation is confined to an examination of the effects of the US Occupational Safety and Health Act (OSHA) of 1970. At best, this research has produced mixed results with respect to OSHA's effectiveness, although more recent research – and, in some ways, methodologically superior – tends to find results that support the hypothesis that direct regulation reduces injury rates. (This question is also taken up by Mendeloff in the Chapter 11 with some comparative discussion on the US and Canada.)

Much of this early research involved a time-series analysis or cross-sectional pre- and post-OSHA comparisons of aggregate injury rate data. By and large, these studies were unable to find the expected reduction in the incidence of workplace injuries (Smith, 1973; Mendeloff, 1979; Currington, 1986). However, Smith (1992, p. 566) notes data problems render such comparisons problematic: "Because the Occupational Safety and Health Act fully covers the private sector, and because before-and-after comparisons are generally infeasible, a convincing study of the overall effects of the Act has not been – and may never be – done".

Another group of studies has evaluated the impact of OSHA enforcement activity – that is, the effect of inspections and fines – on the incidence and severity of workplace injuries. Following Smith (1992), these studies may be classified into two categories: those using aggregate industry injury rate data and those using plant level data.

Enforcement variables used in research examining industry aggregate accident rates include lagged measures of the probability of inspection and the expected penalty for an OSHA violation. In general, these studies found little or no effect for OSHA enforcement activity. For example, Viscusi (1979) was unable to detect a statistically significant relationship between injury rates and either inspection probability or the expected penalty. In a later study, Viscusi (1986) found that OSHA enforcement reduced the lost workday incidence rate by a modest 1.5 to 3.6%, although Smith (1992) argues that this result may have been a statistical artifact – the product of changes in employer reporting behavior resulting from a change in OSHA inspection strategies.

Arguing that these lagged penalty data were as much a measure of employer non-compliance as a proxy for a deterrent effect, Bartel and Thomas (1985) estimated a system of structural equations in which the probability of inspection and penalties per inspection were treated as endogenous. They found that while OSHA significantly reduced employer non-compliance, there was little relationship between non-compliance and the lost-time injury rate. However, these authors conclude that OSHA *indirectly* reduced accident rates by placing a greater regulatory burden – in the form of increased inspection probability – on firms with higher injury rates.

A study of industry aggregate injury rates in Quebec by Lanoie (1992) found a statistically significant negative relationship between inspection probability and the lost-time injury rate. However, the likelihood of a workplace health and safety inspection by an officer of the Quebec government was positively associated with injury severity, measured as average number of workdays lost per injury. In addition, Lanoie failed to detect a statistically significant relationship between probability of penalty and either frequency or severity of work injuries.

Research using plant level data have generally reached more optimistic conclusions about OSHA's effectiveness, although these studies have also produced mixed results. Two types of studies have been conducted. Earlier research compared firms that had been inspected early in the year with firms that had been inspected late in the year, hypothesizing that inspection effects should be more evident for the former group of firms than for the latter group (Smith, 1979; McCaffrey, 1983). Using data from 1973 and 1974, Smith found that 1973 inspections reduced injury rates by about 16% while 1974 inspections induced a 5% reduction, although the latter relationship was not statistically different from zero at conventional levels. McCaffrey failed to find a statistically significant effect using data from 1976-77. As Scholz and Gray (1990, p. 299) note, taken together, these results suggest that "the easily accomplished reductions in risk that OSHA inspections could impose may have already been implemented in 1976, leaving more complex issues of risk reduction, less amenable to quick fixes".

As Smith (1992, p. 569) points out, because these early studies lacked data on citations and fines resulting from inspections, they were only able to measure the abatement of injuries following an inspection; as a result, these studies were unable to measure OSHA's "deterrent" effect. Replicating this research, Ruser and Smith (1991) used a measure of inspection probability based on the average inspection frequency for similar firms to estimate the deterrent effect. On the basis of this analysis, they concluded that: "there is virtually no evidence of a deterrence effect" (p. 231).

Interestingly, recent plant-level studies, which use explicit before-and-after comparisons of the same firms, provide evidence for a more sanguine assessment of direct regulation. Using a unique data set that allowed tracking of inspections and penalties for a large sample of individual firms over a seven year period, Scholz and Gray (1990) estimated both the deterrence and abatement effects of OSHA enforcement activity. They found that a 10% increase in enforcement resulted in a 1% reduction in the accident rate, a much larger effect than detected in prior research, although one that the authors describe as "modest" (p. 302). This reduction was primarily due to a "deterrence" effect and, specifically, an increase in the probability of inspection, as opposed to an increase in the average penalty.[9]

Importantly, Scholz and Gray argue that their results indicate that economic models of occupational safety and health regulation, which assume that firms optimize when making safety and health choices, fail to account for the limited information processing capacity of managers.[10] Due to their limited capacity, managers do not optimize, but often engage in "fire-fighting", responding to problems as they become more significant relative to other issues. As evidence, Scholz and Gray find that an unexpected increase in the accident rate in one year will lead to a reduction in injuries in the next, and vice-versa. In addition, they find a lag between OSHA enforcement activity and a change in firm health and safety – a result that they claim is evidence of an organizational learning curve. Ruser (1985) obtained similar results.

Nonetheless, overall the extant evidence suggests that OSHA has, at best, resulted in a modest improvement in workplace health and safety in the US. However, advocates of direct regulation argue that these disappointing results are primarily due to diffident administration and a lack of funding than to a fundamental flaw in this type of regulatory regime. In particular, they point to two problems. First, the process of adopting permanent health and safety standards under OSHA is slow and cumbersome. Governed by the Federal Administrative Procedures Act, the law requires a Notice of Intended Rulemaking and a subsequent proposal, both of which must be published in the Federal Register. This is followed by a Public Hearing and comment period where all interested parties are invited to submit comments, which the agency must consider before promulgating a standard.[11] After they are issued, standards are subject to judicial review, and the Supreme Court has ruled that the agency must provide substantial evidence that the standard is based on a "significant" risk. Second, agencies responsible for administering the Act are substantially underfunded, a problem exacerbated during the Reagan-Bush administrations. Dorman (1996, p. 193) notes that there are more fish and game wardens in the US than occupational safety and health inspectors.

Critics have expressed greater apprehension over the impact of direct regulation on economic productivity (Burton and Chelius, 1997). There is a public perception, shared by some economists, that the proliferation of industrial regulation in the 1960s, particularly with respect to occupational health and safety and environmental protection was responsible for anemic productivity growth since that time. Research is sparse, however. One study, estimating annual total factor productivity for 450 US industries between 1958 and 1978, found that OSHA accounted for around 19% of the productivity slowdown of the 1970s (Gray, 1987). Viscusi (1996) has estimated the cost and benefits of five OSHA regulations and found that for four of these, the costs of the regulation exceed the benefits in terms of lives saved. However, Stone (1997) challenged Viscusi's estimates, claiming he ignored other benefits, such as the reduction in injuries and illnesses. His re-analysis of one of these regulations showed it was in fact efficient, when these other benefits were considered.

Nonetheless, if one assumes that the direct regulation of workplace safety is inefficient public policy, then it is possible that direct regulation could actually result in the deterioration of worker health. Keeney (1994) has argued that a reduction in disposable income due to these regulatory costs can lead to changes in spending on safety and healthcare more generally, greater stress due to job loss, and risky behavior such as increased alcohol and tobacco consumption.

10.2.3 Internal Responsibility System

A principal criticism of direct regulation is that it fails to recognize firm heterogeneity, so that standards appropriate for one firm are likely to be inappropriate for another. In contrast, the internal responsibility system is highly adaptable to the particular circumstances of the firm and is flexible so that it can respond relatively quickly to technological change. The design of safety "standards" is in the hands of the parties themselves – labor and management – who are intimately familiar with plant operation and who are therefore well placed to implement regulatory standards that are effective and efficient. In addition, administrative costs, which are principally borne by employers, at least initially, are relatively low. Enforcement is in the hands of the firm's workforce so that the probability of detecting a violation will be high.

On the other hand, the success of the internal responsibility system is critically dependent on employee bargaining power. It is likely that internal responsibility is less effective in non-union workplaces than in union ones. In addition, unions are political organizations that necessarily respond to the preferences of their memberships. And safety and health are often given a relatively low priority by a rank and file that sometimes appears to be more interested in wages and job security. Furthermore, the internal responsibility system can be used by employees to shirk legitimate work assignments or by labor unions as leverage in collective negotiations with employers. Finally, there are concerns that labor members may lack the expertise, particularly in the realm of occupational health, to either design effective standards or monitor firm compliance.

Unlike either direct regulation or general financial incentives, there is little direct evidence on the efficacy of the internal responsibility system. Most of this research has examined joint health and safety committees (JHSCs) and much of it uses data on Canadian workplaces.[12] By and large, however, the Canadian studies either examine process issues or factors determining the relative effectiveness of JHSCs rather than the question of whether or not they reduce injury rates or otherwise improve worker health compared to workplaces without such committees. In addition, these studies often rely on subjective reports by the participants rather than objective evidence. Nevertheless, some useful information relevant to the question of the effectiveness of the internal responsibility system may be gleaned from this research.

For example, Shannon *et al.* (1992) find that lower accident rates are found in firms where the JHSCs includes a senior manager; where labor members had access to professional expertise; and where the JHSC had a broad mandate rather than a narrow one. Furthermore, Tuohy and Simard (1993) find that JHSCs were more effective in reducing accident rates when the committee had an equal number of labor and management members and where there are well-established operating procedures. In other words, both studies indicate that JHSCs are more effective when employers give them greater resources and support.

Three studies directly examine the issue of whether JHSCs ameliorate workplace safety. Cooke and Gautschi (1981) combined OSHA administrative data with the results of a survey of 113 manufacturing firms in Maine to investigate, among other things, whether joint labor-management safety programs affected firm injury rates. They obtained mixed results, which depended on firm size. Large firms with joint safety programs had lower injury rates than large firms that did not have a joint program. However, this result was statistically significant only for firms with more than 300 employees and only at the 0.10 confidence level. For small firms, the opposite result was found; firms with joint programs had higher injury rates. Boden *et al.* (1984) surveyed 290 large (more than 500 employees) Massachusetts firms but failed to find a relationship between the presence of a joint safety committee and workplace injury or illness rates. Importantly, both of these studies – Cooke and Gautschi (1981) and Boden (1984) *et al.* – use cross-sectional, rather than longitudinal data, and are, therefore, limited in their ability to address the question of whether there is a causal relationship between JHSCs and workplace safety.

On the other hand, the most careful examination of internal responsibility found that the internal responsibility system generally and JHSCs in particular were associated with lower levels of workplace injuries and illness (Lewchuk *et al.*, 1996). This study used administrative data from the Ontario Workers' Compensation Board supplemented with data from two surveys. The authors found that both enactment of internal responsibility legislation and the introduction of JHSCs were negatively and significantly related to the workplace injury rate. Specifically, they find that JHSCs may reduce lost-time claims by as much as 18% relative to similarly situated firms without JHSCs. Importantly, they also found that joint committees were more effective at reducing injury rates in unionized firms than in non-union firms.

The latter result suggests that unions play an important role determining the effectiveness of JHSCs. Similarly, Weil (1991; 1992) has argued that unions improve the effectiveness of direct regulation. Using 1985 OHSA data from the US, he has shown that unions increase inspection probability; inspection intensity, as measured by the duration of inspections per employee; and the scope of the inspection, i.e. whether or not the inspection resulted in a physical examination of the workplace. Weil also found that unions increased the number of citations as well as the severity of the penalties. These results suggest that there may be a synergy between direct regulation and internal responsibility, at least for unionized workplaces.

On the other hand, critics cite anecdotal evidence that shows that unions use regulatory agencies and, in particular, occupational safety and health agencies to enhance their power in organizing campaigns and in collective bargaining (Northrup, 1997). JHSCs would seem to offer similar opportunities for unions to enhance their organizing and collective bargaining outcomes. However, Schurman *et al.* (1998) note that complaint-based inspections in unionized firms result in a higher percentage of violations than similar inspections in non-union firms and argue that this contradicts an interpretation that unions use safety regulation to gain organizing and bargaining advantage.

Hebdon and Hyatt (1998) present conflicting evidence with respect to this issue. They use Ontario data to examine factors influencing the probability of a refusal to do unsafe work or the probability of a health and safety complaint. In general, they found that while the probability of both events was higher where there is a contentious industrial relations environment, they found no evidence of concerted harassment of employers during collective negotiations.

More generally, we might expect that unionization could lead to more optimal health and safety conditions. Workplace health and safety has characteristics of a public good in that consumption is neither rival nor excludable.[13] In addition, free rider problems may prevent unorganized workers from negotiating the optimal provision of safety conditions by the employer. That is, workers will be individually reluctant to reveal preferences because they fear that they will pay the full cost of safety. Employers must therefore rely on information gleaned from the labor market. However, such information necessarily reflects the preferences of workers who are very different than the average worker; these marginal workers are younger and are less likely to have family responsibilities. Among other things, marginal workers are likely to be less concerned about workplace hazards and should be less willing trade-off wages for increased safety.

On the other hand, unions, which are democratic political organizations, are more likely to reflect the preferences of the average workers. In fact there is some evidence that unions in fact respond to the safety objectives of more senior workers while management is more likely to be influenced by the preferences of marginal workers (Kahn 1987; 1990).

10.2.4 General Financial Incentives

Both direct regulation and internal responsibility attempt to regulate the safety *process*, imposing sanctions on employer *behaviors* thought to affect the accident rate. In contrast, a regime using general financial incentives regulates safety *outcomes*, imposing sanctions based on employer performance with respect to results-based workplace safety measures. One proposal for general financial incentives is the injury tax, whereby the government imposes a monetary penalty for each work-related injury or illness (Smith, 1974). A more prosaic form of general financial incentives is experience-rated workers' compensation insurance, as discussed in the previous section, whereby the firm's compensation assessment is based, wholly or partially on its accident experience.

Like the internal responsibility system, a system of general financial incentives imposes no specific requirement vis-à-vis firm health and safety practices, allowing firms to select the most appropriate means for attaining its safety goals. Furthermore, under a system of general financial incentives, administrative costs will be lower than those incurred under either direct regulation or the internal responsibility system. However, because experience-rating adjustments to workers' compensation assessments are based on the firm's *claim* experience rather than its *accident* experience, experience rating provides employers with incentives to engage in claims management as well as accident prevention. Claims management includes a number of less than desirable practices, including retaliation against workers who initiate compensation claims and legal challenges to legitimate claims by injured workers. In addition, for actuarial reasons, true experience rating is not feasible for small firms.

There is substantial research investigating the impact of experience rating on the frequency and severity of work accidents. In general, these studies have found considerable evidence that experience rating is associated with lower injury rates, although there are a few exceptions (Hyatt and Thomason, 1998). However, research that fails to find the expected effect is, in general, methodologically weaker than studies that do (Hyatt and Thomason, 1998). Studies investigating injury severity have generally produced mixed results. There are two possible explanations for the disappointing results with respect to injury severity: either employers have less ability to affect severity than the incidence of injuries or the effects of experience rating on incidence overwhelm the severity effect. That is, on the margin, experience rating induces employers to reduce the frequency of less severe injuries. In either event, severity studies are generally less informative and will not be reviewed here. However, a brief review of injury rate research follows.

Research examining the impact of experience rating on workplace safety, most of which uses US data, falls into one of three categories. The earliest studies exploited the fact that US experience-rating formulae are different for large and small firms, so that large firms are more likely to be experience rated and are more extensively experience rated than small firms. Since a difference in injury rates between large and small firms could be ascribed to firm size effects unrelated to experience rating – such as, scale economies in accident prevention efforts – these studies examined the relationship between benefit generosity and accident rate. As indicated previously, empirical research conclusively demonstrated work injuries are positively related to benefit levels. However, if experience-rating induces firms to improve workplace safety, then this relationship should be attenuated in large firms relative to small ones. That is, as benefit levels become more generous, experience rated firms will increase their safety investment, partially offsetting the increased level of injuries resulting from worker moral hazard. Several studies found this hypothesized relationship (Ruser, 1985; Butler and Worrall, 1988; Ruser, 1991); only one failed to do so (Chelius and Smith, 1983).

As indicated, the positive relationship between benefit levels and the work injury rate is primarily attributable to a reporting effect; workers are more likely to report an injury when benefit levels are high than when they are low. It is unlikely that fatal claims are subject to this reporting phenomenon, so that the relationship between benefit levels and fatal injury probability should more accurately reflect the impact of benefits on employer behavior. Four studies have examined this relationship, and three found that the incidence of fatal injuries was negatively associated with higher benefit levels, as expected if experience rating has safety-enhancing effects (Moore and Viscusi, 1989; Ruser, 1991; and Durbin and Butler, 1998). Only Butler (1983) failed to find the hypothesized negative relationship.

Several studies have taken advantages of "natural experiments" to compare injury rates before and after the implementation of an experience rating program. Chelius and Kavanaugh (1988) examined injury rates of two New Jersey colleges before and after they elected to self-insure and ceased to be covered by private compensation insurance.[14] Chelius and Smith (1993) compared occupational injury rates for small firms in Washington, which gives experience-rated discounts to these firms, with injury rates for small firms in states that do not offer these discounts. Bruce and Atkins (1993) and Hyatt and Thomason (1998) examined workers' compensation claim rates in Ontario and BC, respectively, before and after the introduction of experience rating in those provinces. Shields *et al.* (1997) explored the effect of the implementation of "large-deductible" compensation insurance policies – where insured firms are responsible for the first several thousand dollars of compensation costs – in Texas. Finally, Durbin and Butler (1998) used state-level US data to investigate the effects of both large and small deductible policies as well as a rule change that lowered eligibility criteria for experience rating.[15] With the sole exception of Chelius and Smith (1993), these experiments found that experience rating was associated with lower injury rates.

Out of 14 studies reviewed here, 11 found evidence that experience rating results in an amelioration of workplace health and safety. This evidence was produced by research that is remarkably mixed with respect to both data sources and methodology. And, as indicated, a careful examination reveals that studies failing to detect this relationship were methodologically weaker than those that did. Taken as a whole the evidence is quite compelling: experience-rating works.

However, as Hyatt and Thomason (1998) point out, the leap from the observation that experience rating is associated with lower injury or claims rates to the conclusion that experience rate enhances firm safety is short, but perilous. Experience rating may lead to increased claims management by employers, including the denial of legitimate compensation claims and threats against workers who file claims, as well as pro-active staffing practices designed to screen job applicants likely to file a workers' compensation claim. This has the effect of reducing injury reporting, while leaving workplace hazards undisturbed. Two studies show experience rating increases employers claims management activity.

Examining a large set of administrative records from Ontario, Hyatt and Kralj (1995) found that experience-rated employers were significantly more likely to appeal claims than non-experience rated employers, and that the likelihood of an appeal for experience-rated employers increased as a function of the financial incentives that they faced. Kralj (1994) analyzed a small survey of Ontario employers in which managers were asked to report their impressions of the effects of experience rating on their behavior, i.e. changes in accident prevention and claims management practices resulting from experience rating. He found that while both prevention and claims management behaviors increased, experience rating had a greater impact on accident prevention efforts. Thus, while it is clear that experience-rating leads to more intensive claims management efforts, this is not the only effect. Furthermore, claims management is not an unalloyed evil. The denial of fraudulent claims is both equitable and efficient, and there is evidence indicating that a prompt return to work leads to more successful rehabilitation.

Using a survey data set consisting of over 450 Quebec manufacturers, Thomason and Pozzebon (forthcoming) examined the estimated relationship between experience rating and a wide range of firm health and safety and claims management practices. These practices included, for example, the amount of health and safety training provided to workers, the extent to which the firm disputed workers' compensation claims, the number of in-house personnel devoted to claims management or accident prevention activities, and firm expenditures on personal protective equipment. They found that experience-rated firms were both more likely to engage in more aggressive claims management and to make greater effort to increase workplace health and safety. Interestingly, however, the evidence also suggested that high wage firms are more likely to reduce workers' compensation claim costs by increasing their accident prevention efforts (relative to their claims management efforts) than low wage firms. This result implies that there may be a "high road" and a "low road" response to experience rating.

10.3 CONCLUSIONS

The past 20 years have seen a substantial accumulation of knowledge concerning the effects of various policy options, although much is left to be learned. It is by no means certain that policy makers have fully taken advantage of this knowledge or that they have developed a coherent policy with respect to occupational health and safety problems. Rather, policy has developed in a piecemeal fashion as jurisdictions have experimented with various approaches to these problems.

Until recently, these approaches tended to emphasize direct regulation and, more recently, internal responsibility; general financial incentives are little used. Workers' compensation programs have only recently introduced experience rating to the assessment process, and in most provinces in Canada, there are restrictions on its application, which substantially limit its effectiveness. For example, in British Columbia the experience rating adjustment is limited to 30% of the base assessment rate.

However, considerable evidence indicates that general financial incentives are effective in reducing accident rates. Moreover, experience-rating does not share many of the problems associated with the other two approaches. In addition, both the costs of direct regulation and its apparent limited effectiveness call into question whether a broad application of direct regulation is appropriate.

Nonetheless, general financial incentives, particularly in the form of experience-rated compensation assessments, are not a panacea. Two problems may be identified. First, because accidents are, by definition, random events, general financial incentives are not easily applied to small firms – the small firm's experience is not necessarily indicative of its underlying safety. Second, due to the long latency of many occupational diseases, it is difficult to assign responsibility to a particular employer. Finally, direct financial incentives assume that firms engage in an optimizing cost benefit calculus, but the evidence suggests that limited information processing capacity may lead managers to satisfice. Under these circumstances, direct regulation could provide a needed shock to focus managerial attention on safety and health problems.

While this implies a continued role for direct regulation, it also suggests a more limited and targeted approach. More specifically, due to the high costs of direct regulation, the resources required by this option should be directed at high-risk industries. This would include, in particular those in which there are numerous small firms, such as construction and those in which there is a relatively high probability of catastrophe – that is, an accident in which there is significant loss of life – such as underground mining. In addition, these resources should also be directed to the problem of long latency occupational disease, where it is unlikely that general financial incentives will be effective. This includes funding research that would investigate the relationship between occupational exposures and subsequent disease development as well as funding for monitoring workplace exposure.

10.4 KEY MESSAGES

- Much has been learned in the last two decades regarding effective policies to reduce disabling injury at work.

- Both direct regulation and internal responsibility have been widely used in Canada, whereas general financial incentives recently have become more pervasive as they are in the US.

- Financial incentives do appear effective in reaching injury rates whereas the limited effectiveness of direct regulation raises questions about its value except where it may be targeted at high-risk individuals and longer-latency occupational disease exposures.

- General financial incentives are limited in their value for small firms.

- A stronger role for workplace exposure surveillance is necessary.

10.5 NOTES

1) Accident prevention costs are manifested in higher production costs and lost productivity, which means that there are fewer goods and services including, for example, medical and rehabilitation services for those claimants who are injured or become ill due to a workplace accident or exposure.

2) The $10,000 figure for accident costs subsumes an evaluation of the monetary value for intangibles such as pain and suffering.

3) This example assumes that workers are risk neutral, i.e. they are indifferent between income that will be paid with certainty (such as the wage income available from safe employers, where there is no risk of injury) and income that is uncertain (such as the wage income paid by hazardous employers, where there is a 10% chance that the worker will lose wage income due to a work accident). If the worker is risk averse, as is generally thought to be the case, then the worker would demand a salary higher than $41,000 to compensate him or her for the additional risk incurred by working for a hazardous employer.

4) It is important to note that workers will continue to have an incentive to avoid workplace accidents and illnesses even if the wage is fully compensating, i.e. if it compensates workers for all of the expected cost of injury. This is because the worker will continue to incur costs if an accident occurs, unless he or she is able to purchase insurance that covers those costs.

5) Interestingly, one of the few studies that failed to find this relationship used a Canadian data set (Martinello and Meng, 1992).

6) It is not feasible to experience rate small firms, i.e. firms with only a few employees. Because work accidents are random events and because their employment base is small, the number of accidents does not provide a reliable estimate of the underlying risk of injury.

7) It is also possible that Occupational Health and Safety regulations could attempt to influence the behavior of employees, although none of the existing regulatory models contemplates this.

8) Alternatively, sanctions raise the costs of not engaging in accident prevention.

9) Scholz and Gray also used their data set to replicate prior research in order to determine reasons for the discrepancy between their results and the results of these earlier analyses. They concluded that Smith (1979) and McCaffrey (1983) failed to find significant abatement effects because they had not accounted for long-term enforcement effects. Smaller deterrence effects found by Viscusi (1986) were attributed to sample differences. Specifically, the Scholz and Gray sample contained plants that were larger, more dangerous, and more heavily inspected than the average manufacturing plant examined by Viscusi. Scholz and Gray hypothesized that the plants in their sample were more amenable to the ameliorative effects of OSHA enforcement than the average plant.

10) The phrase "limited information processing capacity" is not meant to apply only to the abilities (or limitations) of managers. Rather it refers to limitations that constrain us all (including university professors).

11) For example, Meisenhelter (1991) notes that a period of six years elapsed between OSHA's initial work on a Hazards Communication standard – similar to WHMIS – before it was finally issued in November 1983. Over 200 written comments were submitted totaling over 12,000 pages. There were 19 days of hearings, which produced 4,250 pages of transcripts.

12) A literature search uncovered only two studies examining the effect of an aspect of internal responsibility other than JHSCs on workplace safety. Lanoie (1992) estimated the impact of refusals to do unsafe work in Quebec, using industry aggregate data. He failed to find a relationship between the number of refusals per employee and the lost-time injury rate. However, Lanoie's data show that refusals are negatively related to injury severity, although this relationship is only marginally significant in two of his four specifications and statistically not different to zero in the other two. On the other hand, Cousineau et al. (1995), who also used Quebec data, found that refusals were positively related to one type of injury ("struck by or striking against") thought to be particularly susceptible to safety regulation, while negatively related to two other types ("caught in or between" and "falls or slips"). However, the latter two relationships were not statistically significant.

13) A rival good is one that may be consumed by one and only one person. If it is possible to prevent the consumption of a good, it is excludable. A is a good that is both rival and excludable, while, clean air is both non-rival and non-excludable. These distinctions are important because it is generally thought

that a private market is perfectly capable of efficiently providing rival and excludable goods, but not goods that are non-rival and non-excludable.

14) Firms that self insure, an option available in most US states for firms that meet certain fiscal requirements, are, by definition, perfectly experience-rated.

15) In most US states, there is a minimum payroll requirement that a firm must satisfy in order to become experience rated.

10.6 REFERENCES

Bartel, A.P. and Thomas, LG., 1985, Direct and indirect effects of regulation: a new look at OSHA's Impact. *Journal of Law and Economics*, **28**, pp. 1-25.

Boden, L., Hall, J., Levenstein, C. and Punnett, L. 1984, The impact of health and safety committees. *Journal of Occupational Medicine*, **26**, pp. 829-834.

Bruce, C.J. and Atkins, F.J., 1993. Efficiency effects of premium-setting regimes under workers' compensation: Canada and the United States. *Journal of Labor Economics*, **11**, pp. S38-S69.

Burton, J.F, Jr. and Chelius, J.R., 1997, Workplace safety and health regulations: rationale and results. In *Government Regulation of the Employment Relationship*, edited by Bruce E. Kaufman, Madison, WI: IRRA, pp. 253-293.

Butler, R., J., 1983, Wage and injury rate response to shifting levels of workers' compensation. In *Safety and the Work Force*, edited by Worrall, J. D. (Ithaca, NY: ILR Press).

Butler, R., J and. Worrall, J.D., 1983, Workers' compensation: benefits and injury claims rates in the seventies. *Review of Economics and Statistics*, **65**, pp. 580-589.

Butler, R.J. and. Worrall, J.D., 1988, Labor market theory and the distribution of workers' compensation losses. In *Workers' Compensation Insurance Pricing*, edited by Appel, D. and Borba, P.S. (Boston: Kluwer Academic Publishers), pp. 19-34.

Chelius, J.R., 1976, Liability for industrial accidents: a comparison of negligence and strict liability systems. *Journal of Legal Studies*, **5**, pp. 293-325.

Chelius, J.R., 1982, The influence of workers' compensation on safety incentives. *Industrial and Labor Relations Review*, **35**, pp. 235-242.

Chelius, J.R and Kavanaugh, K., 1988, Workers' compensation and the level of occupational injuries. *Journal of Risk and Insurance*, **55**, pp. 315-323.

Chelius, J.R and Smith, R.S., 1983, Experience-rating and injury prevention. In *Safety and the Work Force*, edited by Worrall, J.D. (Ithaca, NY: ILR Press).

Chelius, J.R and Smith, R.S., 1993, The impact of experience rating on employer behaviour: the case of Washington State. In *Workers' Compensation Insurance: Claim Costs, Prices, and Regulation*, Boston, edited by Durbin, D. and Borba, P.S., (Boston: Kluwer Academic Publishers).

Cooke W.N. and Gautschi, F.H. III, 1981. OSHA, plant safety programs, and injury reduction. *Industrial Relations*, **20**, pp. 245-257.

Cousineau, JM., Girard, S. and Lanoie, P., 1995 Safety regulation and specific injury types in Quebec. In *Research in Canadian Workers' Compensation*, edited by Thomason, T. and Chaykowski, R.P., (Kingston, ON: IRC Press).

Currington, W.P., 1986, Safety regulation and workplace injuries. *Southern Economic Journal*, **53**, pp.51-72.

Dickens, W, 1984, Differences between risk premiums in union and nonunion wages and the case for occupational safety regulation. *American Economic Review*, **74**, pp. 320-323.

Dorman, P., 1996, *Markets and Mortality: Economics, Dangerous Work and the Value of Human Life*, (London: Cambridge University Press).

Dorman, P. and Hagstrom, P., 1998, Wage compensation for dangerous work revisited. *Industrial and Labor Relations Review*, **52**, pp. 116-135.

Durbin, D. and Butler, R.J., 1998, Prevention of disability from work related sources. In *New Approaches to Work Disability,* edited by Thomason, T., Burton, J.F. Jr. and Douglas, E.H., (Madison, WI: Industrial Relations Research Association).

Fairris, D., 1992, Compensating payments and hazardous work in union and nonunion settings. *Journal of Labor Research*, **13**, pp. 205-221.

Fishback, P.V., 1987, Liability rules and accident prevention in the workplace: empirical evidence from the early twentieth century, *Journal of Legal Studies*, **16**, pp. 305-329.

Gray, W.B., 1987, The cost of regulation: OSHA, EPA, and the Productivity Slowdown. *American Economic Review*, **77**, pp. 998-1006.

Hebdon, R. and Hyatt, D., 1998, The effects of industrial relations factors on health and safety conflict *Industrial and Labor Relations Review*, **51**, pp. 579-593.

Heinrich, H.W., Peterson, D. and Roos, N., 1980, *Industrial Accident Prevention: A Safety Management Approach*. (New York: McGraw-Hill).

Hirsch, B.T., MacPherson, D.A. and Dumond, J.M., 1997, Workers' compensation recipiency in union and nonunion workplaces. *Industrial and Labor Relations Review*, **50**, pp. 213-236.

Hyatt, D. and Kralj, B, 1995, The impact of workers' compensation experience rating on employer appeals activity. *Industrial Relations*, **34**, pp. 95-106.

Hyatt, D.E. and Thomason, T., 1998, *Evidence on the efficacy of experience rating in British Columbia.* A report prepared for the Royal Commission on Workers' Compensation in British Columbia, unpublished manuscript.

Kahn, S, 1987, Occupational safety and worker preferences: is there a marginal worker. *Review of Economics and Statistics*, **69**, pp. 262-268.

Kahn, S., 1990, What occupational safety tells us about political power in union firms. *Rand Journal of Economics*, **21**, pp. 483-496.

Keeney, R.L., 1994, Mortality risks induced by the costs of regulations. *Journal of Risk and Uncertainty*, **8**, pp. 95-110.

Kralj, B., 1994, Employer responses to workers' compensation insurance experience rating. *Relations Industrielles*, **42**, pp. 41-61.

Lanoie, P., 1992, The impact of occupational safety and health regulation on the risk of workplace accidents: Quebec, 1983-87. *Journal of Human Resources*, **27**, pp. 643-660.

Lewchuk, W., Robb, A.L. and Walters, V., 1996, The effectiveness of Bill 70 and joint health and safety committees in reducing injuries in the workplace: the case of Ontario. *Canadian Public Policy*, **22**, pp. 225-243.

Martinello, F. and Meng, R., 1992, Workplace risks and the value of hazard avoidance. *Canadian Journal of Economics*, **25**, pp. 333-345.

McCaffrey, D., 1983, An assessment of OSHA's recent effects on injury rates. *Journal of Human Resources*, **18**, pp. 131-146.

Meisenhelter, M.C., 1991. The making of OSHA's hazard communication standard: a study of the effects of the outside influence on policy formulation. In *Pennsylvania Economic Association: Proceedings of the Sixth Annual Meeting, May 23-25, 1991* edited by Long, S.G., (Johnstown: University of Pittsburgh), pp. 360-372.

Mendeloff, J., 1979, *Regulating Safety*, (Cambridge, MA: MIT Press).

Moore, M.J. and Viscusi, W.K., 1989, Promoting safety through workers' compensation: the efficacy and net wage costs of injury insurance. *Rand Journal of Economics,* **20**, pp. 499-515.

Northrup, H.R., 1997, Expanding union power by comprehensive corporate campaigns and manipulation of the regulatory process. In *Government Regulation of the Employment Relationship*, edited by. Kaufman, B.E., (Madison, WI: Industrial Relations Research Association).

Olson, C., 1981, An analysis of wage differentials received by workers in dangerous jobs. *Journal of Human Resources*, **16**, pp. 167-185.

Ruser, J.W., 1985, Workers' compensation insurance, experience rating, and occupational injuries. *Rand Journal of Economics*, **16** , pp. 487-503.

Ruser, J.W., 1991, Workers' compensation and occupational injuries and illnesses. *Journal of Labor Economics*, **9**, pp. 325-350.

Ruser, J.W. and Smith, R.S., 1991, Reestimating OSHA's effects: have the data changed? *Journal of Human Resources*, **26**, pp. 212-235.

Sandy, R. and Elliott, R., 1996, Unions and risk: their impact on the level of compensation for fatal risk. *Economica*, **63**, pp. 291-309.

Scholz, J.T. and. Gray, W.B., 1990. OSHA enforcement and workplace injuries: a behavioral approach to risk assessment. *Journal of Risk and Uncertainty*, **3**, pp. 283-305.

Schurman, S.J., Weil, D., Landsbergis, P. and Israel, B.A., 1998, The role of unions and collective bargaining in preventing work-related disability. In *New Approaches to Work Disability*, edited by Thomason, T., Burton, J.F. Jr. and Hyatt, D.E., (Madison, WI: Industrial Relations Research Association).

Shannon, H.S., Walters, V., Lewchuk, W., Richardson, J.L., Verma, D.K.,. Haines, T. and Moran, L.A., 1992, *Health and Safety Approaches in the Workplace: Research Report and Appendices.* (Toronto: IAPA).

Shields, J., Lu, X. and Oswalt, G., 1997, *Workers' compensation insurance deductibles and their impact on employers' costs*. Unpublished manuscript.

Siebert, W.S. and Wei, X., 1994, Compensating wage differentials for workplace accidents: evidence for union and nonunion workers in the U.K. *Journal of Risk and Uncertainty*, **9**, pp. 61-76.

Smith, R.S., 1973, Intertemporal changes in work injury rates. In *Proceedings of the 25th Annual Meeting of the Industrial Relations Research Association*, (Madison, WI: IRRA), pp. 167-174.

Smith, R.S., 1974, The feasibility of an 'injury tax' approach to occupational safety. *Law and Contemporary Problems*, **38**, pp. 730-744.

Smith, R.S., 1979, The impact of OSHA inspections on manufacturing injury rates. *Journal of Human Resources*, **14**, pp. 145-170.

Smith, R.S., 1992, Have OSHA and workers' compensation made the workplace safer. In *Research Frontiers in Industrial Relations and Human Resources*, edited by Lewin, D., Mitchell, O.S. and Sherer, P.D., (Madison, WI: Industrial Relations Research Association), pp. 557-586.

Stone, R.F., 1997, Correspondence on cost-benefit analysis. *Journal of Economic Perspectives*, 11, pp. 187-88.

Thomason, T. and Hyatt, D., 1997, Workers' compensation costs in Canada, 1961-1993. In *Transition and Structural Change in the North American Labour Market,* edited by Abbott, M., Beach, C. and Chaykowski, R., (Kingston, ONT: Industrial Relations Centre and John Deutsch Institute, Queen's University), pp. 235-255.

Thomason, T. and Pozzebon, S., 1995, The effect of benefits on workers' compensation claims incidence in Canada: a micro-level analysis. In *Research in Canadian Workers' Compensation*, edited by Thomason, T. and Chaykowski, R.P., (Kingston, Ont: IRC Press), pp. 53-71.

Thomason, T. and Pozzebon, S., 2002, The determinants of firm health and safety and claims management practices. *Industrial and Labor Relations Review*, 55 (2), p. 286

Tuohy, C. and Simard, M., 1993, The impact of joint health and safety committees in Ontario and Quebec. Prepared for *The Canadian Association of Administrators of Labour Law.* (Toronto: Ontario Ministry of Labour).

Viscusi, W. K., 1979, The impact of occupational safety and health regulation. *Bell Journal of Economics*, 10, pp. 117-140.

Viscusi, W. K., 1986, The impact of occupational safety and health regulation, 1979-83. *Rand Journal of Economics*, 17, pp. 567-580.

Viscusi, W. K., 1993, The value of risks to life and health. *Journal of Economic Literature*, 31, pp. 1912-1946.

Viscusi, W. K., 1996, Economic foundation of the current regulatory reform efforts. *Journal of Economic Perspectives*, 10, pp. 119-134.

Weil, D., 1991, Enforcing OSHA: the role of labor unions. *Industrial Relations*, 30, pp. 20-36.

Weil, D., 1992, Building safety: the role of construction unions in the enforcement of OSHA, *Journal of Labor Research*, 13, pp. 121-132.

CHAPTER 11

Regulatory Approaches to Preventing Workplace Injury

John Mendeloff

This paper has three objectives. It reviews what is known about the effects of regulation on preventing occupational injuries and illnesses; identifies some of the important questions that we do not have answers to, and examines some of the barriers to learning the answers and to suggest what needs to be done to overcome them. The term regulation usually refers to policies that operate through establishing and enforcing rules for the behavior of private or governmental actors, as opposed to policies that operate through the public provision of goods and services or though taxes and transfers. Some attention is also given here to alternatives to regulation, like government programs that offer free advice to workplaces about how to reduce hazards. The role of workers' compensation programs in prevention is dealt with in Chapter 10 and elsewhere in this book. The geographic scope of this chapter includes the United States and Canada.[1]

11.1 PRELIMINARY ISSUES

11.1.1 Theory: Why Should We Expect Regulation to Have an Impact?

Regulations may influence behavior to the extent that they:

1. effectively alter the rights and responsibilities of different parties
2. alter the information on which they make choices
3. change the costs and benefits of the choices.

Most discussions of regulation effectiveness focus on the third point. Deterring non-compliance with regulations is assumed to be a function of the probability of punishment and the size of the punishment. "General deterrence" refers to the effect on all actors, whether they are inspected or not. "Specific deterrence" refers to the impact on those who are actually inspected. Firms may comply with rules because their own calculus, even ignoring the threat of punishment, convinces them that the benefits of compliance outweigh the costs. In some cases regulators may provide information that changes this calculus. In such cases, it is the provision of information, not regulation *per se*, that affects behavior, although the regulation may have made the information more salient.

Some have noted that the limited resources of regulatory agencies like the federal Occupational Safety and Health Administration (OSHA) prevent it from inspecting more than a few per cent of all workplaces each year, even in sectors like manufacturing. Often overlooked is the potential role of worker complaints; a complaint can raise the probability of an inspection from almost zero to 100%. Thus a key feature of any system, especially those with small inspectorates, is the role that workers play in monitoring. Once a firm is cited for a violation, current American rules usually make it irrational not to comply; however, other systems are much more tolerant of foot-dragging. The long-run impact of citing violations may depend significantly on the average duration of compliance. How often do compliance measures eliminate a problem for ten or 20 years? How often do they quickly break down?

Apart from the impact of deterrence on them, firms may adopt a general policy of obeying laws simply because they believe they have a responsibility to do so. Exceptions may be made for rules that seem especially unreasonable or costly. Some observers (Ayres and Braithwaite, 1994) have argued that regulatory agencies that treat all firms as potential miscreants rather than as potential collaborators may alienate those that are, in fact, largely law-abiding, and drive them into a more adversarial stance.

Regulation can also provide rights to workers – e.g. to complain to authorities without fear of retribution, to participate in plant level safety decision-making, to refuse assignments of dangerous work – that may enhance the level of safety. In concept, regulators could also choose to punish workers who violate safety rules; however, this approach has generally been eschewed, leaving the issue of worker discipline in the hands of the employer.

From the perspective of regulatory agencies, the most effective strategy to pursue should depend partly upon the scope of the hazard problem it is able to address. For example, occupational safety agencies whose standards address only a minority of all injuries have good reason to think about how to leverage interventions to encourage firms to deal with hazards outside of the regulatory scope.

11.1.2 Accuracy of Injury and Illness Data

Unfortunately, figures on injury and illness rates do not always give us a true picture of the level of risk at workplaces. The shortcomings in the reporting of occupational illnesses, especially those with long latency periods, are well known. (US Department of Labor, 1994) In the US, the OSHAs own inspections have often found numerous instances of injury under-reporting (National Research Council, 1987). The reports of the duration of disability in the BLS Survey have also been found to be underestimates (Oleinick *et al.*, 1993). Other studies have indicated that musculoskeletal disorders (MSDs) may be substantially under-reported in the US (Fine *et al.*, 1984). If under-reporting were constant across firms and across time, it would not pose a major problem for evaluations. However, there are many examples of changes or differences in reporting, including the following:

1. Many studies have shown that the number of reported non-fatal injuries increased when workers' compensation (WC) benefits increased (see Chapter 10 by Thomason). Apparently, whatever incentives higher benefits provide to employers either to prevent injuries or to fight claims have, on average, been outweighed by the incentive to report that they provide to workers. [2]
2. Fatalities are much less subject to reporting incentives, but there is evidence that the probability that survivors will apply for death benefits is affected by the level of death benefits, at least for cases where the workplace etiology is uncertain (Smith, 1992).
3. The drop in reporting of non-lost-workday cases in the early 1970s in the US was clearly a reporting artifact, reflecting increased employer understanding of what this category included.
4. When OSHA targeted workplaces in the early 1980s on the basis of their lost workday injury rates, reporting among the affected firms declined by an average of 5 to 10% according to studies by Ruser and Smith (1988; 1991)
5. More recently, Waehrer *et al.* (1997) studied the recent growth in the percentage of lost workday cases in the US that involved only "restricted work activity" (RWA) and not actual days away from work (DAW). They concluded that "...higher income replacement rates and shorter state waiting periods increase the provision of restricted workdays which in turn substitute for days away from work for sprains and strains and carpal tunnel cases".
6. Some states require that employers send in reports about non-lost-workday cases, while others do not. Mendeloff (1996) found that, in the construction industry, establishments in the former group of states reported non-lost-workday cases to the Bureau of Labor Statistics's (BLS) Survey of Occupational Injuries and Illnesses at a rate 15% higher than did those in the latter group of states.
7. In the BLS Survey, establishments with fewer than 20 workers have much lower lost workday injury rates than establishments with 20 to 250 workers. In contrast, their fatality rates are far higher than those in larger establishments. Although several explanations are possible, the most likely appears to be that under-reporting of less severe injuries is greater in the smallest workplaces (Mendeloff, 1996b).

These examples do not necessarily preclude the usefulness of injury data for evaluation, but they do show the need for caution in assuming that reported figures are accurate measures of risk.

11.1.3 Generalizability of Findings

It is also important to keep in mind that findings from any particular study may not be generalizable over time or across jurisdictions. Different findings might arise because of several factors: changes in the actual policy and practices of regulators, changes in the environment facing firms, or differences in the sample of firms examined. An example of the first factor is that typical OSHA inspections in 1974, 1983, and 1998 all differed in potentially significant respects. In the first year, federal OSHA conducted 80,000 inspections, only 2% of the violations it cited were classified as "serious," and average penalties per inspection were approximately $100 (Siskind, 1993). In the second year, OSHA followed a "records-check" policy in manufacturing that required inspectors to leave without actually examining the worksite if the establishment had an injury and illness rate below the manufacturing average. In the third year (1998), close to 50% of manufacturing inspections were levying penalties, a bigger share of inspections took place at workplaces with more than 100 workers, most violations were cited as "serious", and the average fine per inspection was several thousand dollars. Million dollar fines had become a regular, if still unusual, occurrence, but the number of total inspections had fallen to about 30,000 per year. While it is unclear whether these and other variations make a difference, we should be cautious about generalizing from results of any one study. Finally, of course, different findings among studies may also arise because of differences in methodology which can affect their internal validity.

11.2 WHAT WE KNOW

11.2.1 Inspections and Injuries

Approximately 20 studies have examined the effects of OSHA inspections; most used American data, although a few studies looked at Canada. Thomason (see Chapter 10) provides a review of many of these studies; Smith (1992) provides the most complete review and the most extensive methodological critique. Here, we will focus only on some highlights.

1. Both safety and health inspections in manufacturing have led to decreases in the level of non-compliance with standards at the inspected workplace (Weil, 1996; Gray and Jones, 1991).
2. The impact of inspections on compliance in the construction industry, although positive, was smaller than in manufacturing (Weil, 2000).
3. More inspections in an industry do lead to decreases in the level of non-compliance with standards in that industry (Bartel and Thomas, 1985).
4. Despite the effects of inspections on compliance, many studies failed to find a clear impact of inspections on injuries. One factor contributing to that failure was that probably only a minority of injuries are caused by violations of specific OSHA standards and a large percentage of those were momentary and unlikely to be detected by inspections (Mendeloff 1979; 1984). The first of these conclusions could, as discussed below, be altered by the adoption of standards addressing ergonomic hazards and requiring "comprehensive safety and health programs". However, while these rules increase the scope of enforcement programs, their actual impact has yet to be clearly demonstrated.

More generally, earlier studies of inspection impact had several limitations:

a) Many were limited to aggregate industry data relating inspections to injury outcomes.

b)   Most studies that used individual establishment data lacked any information about the characteristics of the inspection; the few exceptions used small samples.

c)   They looked at impacts only in the year immediately following the inspection.

It was not until 1990 that Gray and Scholz (1990) overcame most of these problems by linking OSHA inspection data with establishment-level injury rates from the BLS Survey of Occupational Injuries and Illnesses. They found that inspections with penalties reduced the number of injuries by 15 to 22% in the two years following federal OSHA inspections in manufacturing from 1979-85 (BLS, 1993). In the data set they used, the presence of a penalty was essentially equivalent to the citation of a "serious" violation. Inspections which did not result in penalties had no impact or perhaps even a positive impact. Their data set included 6,842 establishments with seven continuous years of BLS Survey data in the 28 states where federal OSHA operated the enforcement program. The mean size of the establishments was some 450, much larger than the average manufacturing establishment.

Later, Gray (1996) replicated the analysis for BLS data from 1987-91. This time the mean size of the establishments was over 650 for those with five consecutive years of injury data; and about 250 for those with two to four years of data. His findings were similar in many respects, although the estimate of impact dropped to some 15%, still limited to those inspections which levied penalties. One somewhat disturbing feature of this study was that not only did the lagged effects of the inspection continue into the third year after the inspection, but the third year effects constituted at least one-half of the total impact. While it is reasonable that the effects of inspections on injuries may accrue over several years, it seems less plausible that the bulk of the impact would be felt three years after the inspection

Still, it seems reasonable to conclude the following.

1.   In large plants in the US manufacturing sector from 1979-91, inspections which levied penalties were followed by reductions in lost workday injuries of 15% to 20% over the two or three years following the inspection.

2.   The size of the penalty did not appear to have any impact on the size of the injury impact.

3.   The smallest effects were at plants with more than 500 workers.

Gray and Mendeloff (2002) recently replicated these studies for the years 1992 to 98 and found that the average effects of federal OSHA inspections at inspected establishments had disappeared. An analysis of four different size classes of establishments found effects only in the 100-249 size group. (In the states that ran their own programs under federal oversight, which had not been examined before, the only effects were found in the establishments with fewer than 100 workers.) The reasons for this erosion are not clear. In concept, a drop in specific deterrence could be explained by an increase in general deterrence. Big fines and a very high probability of inspection and detection will create such strong pressures to comply ("general deterrence") that the added effect of an actual inspection, "specific deterrence", may be quite small. But compared to earlier periods, the number of inspections in 1992-98 was down sharply. And, although the average size of penalties was up dramatically from the 1979-85 period, it had actually declined from 1987-91. Therefore, it is hard to argue that the drop in specific deterrence could have been due to a major increase in general deterrence.

Another possible explanation is declining marginal effectiveness – over time, additional inspections at the same plant may have less and less added effect. While this theory is plausible and consistent with the weaker effects at the larger (and more frequently inspected) workplaces, Gray and Mendeloff did not find that the effect of an inspection increased with the length of time since the last inspection. In addition, it is hard to imagine that the decline envisioned in this theory could have happened so rapidly, i.e. from 1987-91, to 1992-98.

One argument presented by the earlier Scholz-Gray studies was that the impact of inspections might depend at least as much, if not more, on managements being "surprised" by the negative inspection findings as on the actual correction of the specific hazards cited in the inspection. In 1992 to 98, penalties would have been less surprising to managers at the establishments in the sample since over 75% of inspections imposed them. This figure was up from 35% in 1979 to 1987 and 52% in 1987 to 1991. Since managers can now assume that a penalty is likely if an inspection occurs, the inspection may carry less ability to shock managers into action. If so, then the impact of inspections would now have to depend more heavily than before on the mechanism of detection and correction, and less on shock to management.

The previous explanation might be linked to another: that, over time, the percentage of injuries, never very large, that are related to OSHA's standards has shrunk due to changing technologies. In addition, injury rates were declining through most of the 1990s. One reason for this reduction was probably the adoption of more sophisticated disability management programs, which often encouraged workers to return to work earlier. In the United States, at least, an increase in the rate at which WC costs were rising may have fostered these changes in firms' practices. Changes in technologies and disability management may both have made the "detection/correction" mechanism less successful than it had been. All of these explanations remain speculative. What is clear is the effect of federal OSHA inspections on reported injuries and illnesses eroded sharply in the 1990s.

11.2.2 Health Inspections and Exposures

Although the number of safety inspections conducted by federal OSHA is much larger than the number of health inspections, the number of health inspectors is not that much smaller, reflecting the much more extensive efforts required by the latter. Despite this near equivalence in resources, less attention has been devoted to studying the impacts of health inspections. Because illness data are too flawed to serve as good outcome measures, some analysts have looked at exposure data from OSHA health inspections.

Gray and Jones (1991) examined federal OSHA health inspection data from 1972 (essentially the beginning of the OSHA program) to 1983. For workplaces with multiple inspections, they found a decreasing number of overexposures with each successive inspection, although most of the decrease (60% to 90%, depending on the model) occurred between the first and second inspections. Thus it appears that in the early years of the OSHA program, an initial inspection finding of overexposures to toxic substances led to improvements and fewer overexposures by the time of the next inspection measuring exposures.

Authors who have used inspection data to look more closely at exposure data for specific substances have come to mixed conclusions. Froines *et al.* (1990) discerned no downward trend in the findings of OSHA inspections where lead exposures were measured. However, Gomez (1997), who used more precise controls (e.g. occupational categories, in addition to controls for industry, inspection type, and establishment size), did find that lead exposures had declined steadily over time. He did not, however, find a trend in exposures for the two other substances he examined, silica and trichloroethylene. (As noted in the discussion of new standards below, other studies have shown that lead exposures did decline substantially after the issuance of an OSHA lead standard in 1978.)

One striking finding is that exposures above the OSHA permissible exposure limit (PEL) are found in only about 2% of health inspections and that only three hazards – noise, lead, and silica – account for more than one-half of the overexposures. The paucity of overexposures reflects, to some degree, the fact that most OSHA PELs are more than 30 years old. However, as we note below in our discussion of standards, compliance with OSHA's handful of new standards, which typically lowered exposure limits by 95%, has also generally been good.

11.2.3 The Impact of New Standards

Many reviews of "OSHA's impact" have ignored the impacts of new standards. If new standards only have an impact on injuries or illnesses when inspectors cite them, then there would be no reason to look beyond the impact of inspections. However, there is evidence that many employers make efforts to comply even in the absence of a citation (Weil, 1996). Often, this compliance may be due to the deterrent threat of inspection. However, often the deterrent is too small to warrant compliance on the basis of minimizing expected costs. Some firms make it a policy to comply with regulations.

Although there have been relatively few good retrospective studies of what really happened after promulgation, the levels of exposure to most (but not all) of the subjects of new health standards have declined sharply. The examples include asbestos, vinyl chloride, lead (in foundries, battery plants, and primary smelters), cotton dust, and ethylene oxide (Mendeloff, 1988; Viscusi, 1985, US Congress OTA, 1995). Nevertheless, it seems likely that OSHA's estimates of the impact of new standards on injuries and illnesses will tend, on average, to be too high. The general incentive for OSHA is to make high estimates; in addition, in the case of some health standards, the risk assessment methods OSHA uses probably lead to overestimates. Other studies have noted that the costs of compliance with new health standards also tend, on average, to be overestimated (US Congress OTA, 1995). Thus it may be that, on average, the estimates of the *cost per unit of health or safety gained* is roughly right, although the evidence on this point is still only suggestive (Harrington *et al.*, 2000).

We reviewed the estimates of effects that OSHA provided in the preambles to the roughly 30 health and safety standards it has promulgated since 1987. OSHA estimated that these standards would prevent 2,625 deaths per year, the majority related to toxic substances. OSHA claimed the respiratory standard would prevent 952 deaths per year; the lead in construction standard, 420; and the bloodborne pathogens standard, 192. Substance-specific standards had much smaller effects, ranging from 34 deaths prevented for the methylene chloride standard to only one each for the formaldehyde and 1,3 butadiene standards.

Leading the death prevention estimates for safety standards (about 800-900 deaths prevented in total) were the process safety management standard (about 200 deaths prevented annually), the lockout-tagout standard (122), logging operations (111), electrical work practices (78), and concrete and masonry (74).

OSHA also claimed that these new standards would prevent about 150,000 lost workday injury cases each year and an equal number of non-lost workday cases. Safety standards accounted for the great majority of these non-fatal effects. In addition, several impacts attributed to the health standards (e.g. preventing kidney dysfunction or elevated blood lead levels) were not placed in any of the injury categories cited above.

As we observed above, firms may adopt safety and health measures for reasons other than OSHA pressure. Thus one important rival explanation to the argument that OSHA's standards cause changes in injuries or exposures is that new information about the hazard spurred both the OSHA standard and new practices at firms. In other words, the changes would have occurred even without the OSHA standard due to the new information. One test of this model examined changes in exposures for chemicals for which the American Conference of Government Industrial Hygienists had recommended lower exposures (Mendeloff, 1988). It found no impact of the recommendations, in contrast to the impacts from new OSHA standards.

Even if the actual effects are only one-half as large as OSHA projected, the effects are still substantial. Currently, approximately 6,000 occupational accident deaths are reported annually in the United States among employees (BLS, 2000). Four hundred injury deaths per year would constitute almost 7% of that total. Half of the 150,000 lost workday injuries constitutes close to 4% of the number of injuries in that category. Because most disease deaths are not identified by existing reporting systems, it is harder to know what percentage the 1,750 deaths projected by OSHA would comprise of all occupational disease deaths.

In November 2000, OSHA issued an ergonomic standard that it claimed would prevent 4.4 million musculoskeletal injuries (MSDs) over the next ten years. Even though this number is surely an overestimate, a figure even one-quarter this size would dwarf the total effects of all previous standards. A few months later, the Congress and new President overturned this regulation, using, for the first time, a new procedure established by the 1995 Small Business and Regulatory Flexibility Act. Opposition centered on the potential (and hotly debated) costs of the standard to employers. Especially contentious was a provision that required employers to remove workers with MSD symptoms from jobs that could aggravate those symptoms and also to temporarily guarantee them no loss in pay or benefits while they were at another job or off work entirely. This sort of "medical removal protection" has been a part of several OSHA health standards (e.g. lead) and has been justified as necessary to get workers to agree to be tested. However, in this case the potential number of workers affected was especially large and there was an absence of relatively objective indicators (e.g. like lead levels in the blood) of the workers' vulnerability.

Designing a regulatory approach to ergonomic problems is difficult. Earlier proposals had called for "risk scorecards" for each worker, based on specific behaviors like the weight a worker had to lift and the angle of the lifting or the number of times that arms had to be raised above the head. These had been criticized as too inflexible. In its 2000 ergonomic rule, OSHA had bent over backwards to be flexible, raising criticisms that neither employers nor compliance officers would be able to understand exactly what the employer had to do to be in compliance. While the Labor Department decides whether to pursue any new regulatory initiatives on ergonomics, or to rely instead on voluntary programs, it would be useful to try to rigorously evaluate the few regulatory initiatives on ergonomics that have been undertaken by the State of Washington and some others. Another large question that goes beyond the scope of this chapter is whether the benefits of new occupational safety and health standards have been greater than their costs. Among the works that have addressed this question are Morrall (1986), Mendeloff (1988), Viscusi (1983; 1985), and Heinzerling (1998).

11.2.4 Inspections and Targeting

In a study examining the impact of OSHA inspections from 1973 to 1983, Viscusi (1986) observed "inspections stemming from worker complaints...are generally ineffective and lead to few violations, perhaps because workers use these inspections to express other work-related grievances". Employers often have anecdotes to support this view. Nevertheless, there are studies that have found that complaint inspections were more likely than planned inspections to cite serious violations (Weil, 1992; Smith, 1986). More striking, in the mid-1990s the workplaces that were the subject of complaints had higher injury rates than the workplaces targeted by programmed inspections (Mendeloff, 1996). This finding was true even though programmed inspections through 1997 in the US were basically targeted at establishments in industries which were in the upper half of the injury rate distribution. Even within detailed "four-digit" industry categories, the establishments with complaint inspections usually had higher injury rates than those with programmed inspections. In addition, the higher the injury rate of an establishment, relative to its industry average, the more likely OSHA was to find serious violations in a complaint inspection. Finally, the evidence on the effects of inspections on injury rates, although it seems to vary a bit from period to period, does not indicate that complaint and programmed inspections differ very much (Gray and Mendeloff, 2002). Thus, while complaints may sometimes be used as tools in labor-management conflicts, there is strong evidence that, in general, they direct inspectors to workplaces with significant injury problems. It is possible this conclusion would be different if the number of complaint inspections returned to the level of the late 1970s, when there were two to three times as many as in subsequent years; or if planned inspections did a better job of targeting the highest rate establishments.

The design of targeting systems and the optimal mix of inspection types is too large an issue to be treated fully here. However, it seems likely that there is no optimal mix, at least not one that is stable over time. Any single targeting algorithm may be good at addressing some problems but not good at addressing others. And all are probably subject to declining marginal effectiveness. To say that none is optimal is, of course, not to deny that some are better than others. It is probably best to use many different kinds of data to target interventions to different problems: data showing which establishments had high rates within their industries, data showing trends in establishment rates over time, data showing which types of injuries were occurring, and so on.

11.2.5 Voluntary and "Semi-voluntary" Programs

One set of critics described OSHA's enforcement program in the 1970s as imposing "flea bites": a combination of many citations for minor hazards along with small penalties was sufficient to annoy employers, but not to change their behavior (Ayres and Braithewaite, 1994). Actually, as we have seen, the only good evidence we have on the impact of OSHA inspections in the 1970s indicates that inspections were linked to sizable drops in exposure to toxic substances. In any event, the policies of the 1970s did not continue. "Other than serious" hazards now rarely carry any fine, while the penalties for more serious breaches have risen substantially. Bardach and Kagan (1983) and Ayres and Braithewaite (1994) also raised some larger questions: does an adversarial style of regulation undermine the good will of those employers (whom they assume are in the great majority) who are inclined to comply and to invest in worker safety. If so, in addition to alienating an important group from government, the effect can be to greatly complicate and gum up the regulatory process. A team of inspectors may spend months in a single large worksite, documenting literally thousands of violations. Then they may spend many more months in hearings dealing with the appeals from that employer.

It was precisely these experiences that impelled the OSHA staff in Maine to search for a different paradigm. In addition, they were troubled by two other factors: first, that so many injuries were not addressed by OSHA standards; and second, that many lower-injury rate sectors of the workforce were virtually untouched by OSHA, even though they accounted for a large proportion of all injuries. In 1993, the staff created the Maine "Top 200" program. The name referred not to the firms with the highest injury rates, but rather to the firms with the largest number of injuries. Thus, in addition to large manufacturing plants, it included universities, hospitals, large retailers, and fast food chains. The 200 firms accounted for 37% of the injuries in the state.

The program was premised on the assumptions underlying management practices like "continuous quality improvement" and "total quality management". From the first, the message was that everyone (not just the "worst") could improve their performance; and that more can be gained by pushing down the mode of the injury rate distribution than by pushing down its upper tail. The second emphasized the importance of focusing on the adequacy of management systems rather than on the negligence or malfeasance of individuals.

Both of these tools tend to lead enforcers away from an adversarial approach. Instead of focusing on "bad apples", they lead to a focus on those with the most potential to do good (i.e. prevent the most injuries). Instead of focusing on non-compliance with standards, they focus on the establishment of good management systems. The Maine staff embraced this new role for OSHA. If firms would commit to putting the elements of good safety and health programs in place (including a commitment to study and reduce ergonomic hazards), OSHA would refrain from inspections (other than those based on worker complaints or catastrophic accidents). If violations were observed, they would be ignored if the firm had already identified them and had a reasonable timetable for their elimination. OSHA redefined its role from looking for and citing violations to ensuring that the safety and health program elements were operating properly.

Was the Maine Top 200 Program successful? It did win almost unalloyed support from employers, while local labor unions' responses were mixed, but not actively antagonistic. The program caught the attention of President Clinton, who heralded it as the Administration's model for regulatory reform. The Ford Foundation conferred one of its Innovation Awards on it. OSHA planned to use it as a model, but this effort failed for two reasons. First, it quickly became clear that the success of the program in Maine depended in large part on the skillful and time-consuming efforts by the staff there to build a consensus among the regulatory stakeholders. Simply transferring the program to new areas without this prior effort would be much more likely to lead to defections by employers and unions, defections that could easily destroy the mutual trust on which the program depended. Second, and more definitively, a judicial decision barred OSHA from requiring firms to choose between committing to a "Top 200"-type program or agreeing to more intensive OSHA inspections.[3] The court ruled that this policy constituted "backdoor rulemaking," because it would coerce firms into adopting programs (e.g. for ergonomics and other elements of a safety and health program) that had never been adopted through rulemaking. But did the Top 200 program prevent injuries? An evaluation commissioned by OSHA (Mendeloff, 1995) found the results inconclusive. There was some indication, although quite weak, that injuries at Top 200 firms had declined; however, even if true, it was plausible that injuries elsewhere increased due to fewer resources being devoted to inspections as more resources were devoted to monitoring the Top 200.[4]

OSHA conducts a number of other voluntary programs, including its Voluntary Protection Program to recognize firms demonstrating high performance. The largest voluntary program and the only one to have been evaluated is consultation, which provides about 23,000 free consultations, mainly to smaller firms in high hazard industries. The evaluation (Gray and Mendeloff, 2002) indicates consultations do improve compliance with standards. There is also some evidence that they reduce injuries; however, the study was not able to adequately control for selection bias (i.e. firms that request consultations may be more motivated to and more likely to improve, even in the absence of a consultation). Perhaps the clearest finding was that the demand for consultations was heavily influenced by the perceived threat of enforcement. Thus this type of voluntary program seems to be a complement rather than a substitute for enforcement.

11.2.6 Characterizing Regulation in Canada

Canadian public policies on workplace safety and health differ from those in the US in several respects (O'Grady, 2000). Canada relies less on experience rating to provide safety incentives to firms to reduce their injury costs. Although the exact components vary from province to province, it relies more on what has been referred to as the "internal responsibility system" (IRS). The IRS includes two major elements: 1) greater protections to workers who refuse hazardous work; and 2) legal requirements for joint labor-management safety committees with specified functions. Evidence on the impact of these programs is mixed, but appears to show benefits under some conditions. Tuohy and Simard (1992) found, in a retrospective study, that certain kinds of joint committees (those with a broad scope and institutionalized methods) were associated with lower injury rates in Ontario and Quebec. In 1994, the Workplace Health and Safety Agency in Ontario observed that lost-work injury claims did not decline after a 1979 law which required joint committees in more hazardous industries. However, Lewchuk *et al.* (1996), in a somewhat more sophisticated study, found the opposite.

Another major difference between Canada and the United States, of course, is the much higher rate of unionization in the former. To the extent that unions are a prerequisite for a strong health and safety committee and for protecting workers' rights to refuse hazardous work, Canada is better equipped to rely on these tools. Finally, it is intriguing to inquire to what extent the Canadian tradition of law-abidingness and respect for authority (at least relative to the American tradition) might lead to greater compliance than found in the United States.

With respect to more traditional enforcement programs, the provinces, which are responsible for occupational safety and health regulation, display some significant differences among themselves. For example, O'Grady (1999) reports that in the late 1990s, Alberta and Manitoba had about twice as many inspectors per 100,000 workers as Quebec. However, even Quebec had nine inspectors per 100,000 workers, while the figure for federal OSHA in the United States in 2001 is less than two. (For states that operate their own programs, the figure is closer to three per 100,000.) The differences in the number of inspections per worker largely mirror these figures on inspectors. In 1999, Ontario conducted 39,000 inspections, along with 15,000 "investigations", which include what, in the United States, would be classified as complaint and accident inspections.[5] Thus, Ontario's 54,000 figure should be compared to federal OSHA's 30,000 inspections and to the 100,000 conducted for the whole United States out of about 5 million establishments, 2% of the total. With a population of 11 million people compared to the 270 million in the US, these numbers suggest an inspection frequency rate more than ten times higher in Ontario. (In both countries, more than one inspection can occur at a workplace, so that the actual percentage of workplaces inspected will be lower than these numbers suggest.)

An even more striking difference appears in the frequency with which penalties are levied. In the 1990s, US authorities issued penalties in 40 to 50% of their manufacturing inspections. Brown's study of British Columbia in 1984-86 (1994) indicated that a penalty was levied in 1 out of 200 inspections, a rate that he suggested exceeded that in most other provinces at that time. In Ontario in 2000, 134 employers were fined, a ratio closer to 1 in 400.[6] Yet those employers were fined an average of $38,000 for a total of over $5 million. In contrast, the fines in America (for all inspections) averaged about $2,500 per company fined, for a total of about $115 million in penalties. The ratio of fines in America and Ontario ($115 to $5 million) is close to the ratio in their populations. Even if we take into account the lower value of the Canadian dollar, it appears that the total fines levied to enforce regulations are roughly similar on a population basis. Even though fines in Ontario can only be levied following a successful prosecution in the regular courts, which surely reduces the incentive to pursue smaller fines, the totals end up being comparable. What differs is the far greater prevalence of fines in America and the greater average size in Ontario.

Ontario also differs from the US in allowing fines to be levied against employees, both supervisors (70 fines in 2000 with an average of $773) and other workers (92 fines in 2000 with an average of $225). Tickets and summonses are often used in these cases. In Ontario, the Ministry of Labor issues an average of just over one "order" per inspection. OSHA issues an average of two serious citations per inspection. Whether this difference indicates a greater degree of non-compliance in the US is unclear.

Do these differences in enforcement practices have any impact on enforcement effectiveness? The absence of penalties for most violations detected in Ontario may inhibit compliance activity prior to any inspections. However, inspections and, therefore, detection of non-compliance are far more frequent in Ontario. And once an "order" has been issued, the prevalence of unions surely helps to ensure that firms comply. To date, empirical studies of inspection effectiveness in Canada have been conducted with data from Quebec. Using data aggregated at the industry level, Lanoie (1992a,b,c) and Cousineau *et al.* (1995) found results similar to the American studies using aggregated industry data: little or no overall impact of inspections. It seems plausible that the high frequency rate of inspections in much of Canada has been effective in reducing injury rates, especially in conjunction with a high rate of unionization and relatively weak safety incentives from workers' compensation.[7] However, declining marginal effectiveness may also make the marginal effect of current inspections fairly small. The general policy recommendation in these cases is to focus more on the rarely inspected establishments (the "external margin"); however, it would probably be useful to carry out studies to find out how the number of violations found varies as a function of the time since the last inspection.

11.2.7 What Would We Like to Know? And Can We Learn It?

In addition to asking, "what we know", it seems at least as important to ask what we would like to know. What information would be potentially useful to regulatory agencies and help them to do their job better? Any listing is somewhat subjective, but I would suggest the following five topics as among the most important:

1   An understanding of the marginal and interaction effects of various interventions – inspections, training, consultation. An important question here concerns the relative role of cooperative and adversarial approaches.

2   An understanding of what the different elements of employer occupational
     safety and health programs contribute.
3   What more, if anything, needs to be done to ensure that new, potentially
     major occupational health hazards do not emerge? In other words, how
     do we prevent the "next asbestos"?
4   To what extent can regulation help to prevent chronic musculoskeletal
     disorders (MSDs) and how can it be done?
5   Is work-related stress an appropriate focus for government regulation?

In the discussion that follows, I will focus on the first two questions, along
with some attention to the third. The last two questions are left to others to address.

11.2.8 Understanding the Marginal Effects of Different Interventions and Combinations of Interventions

As we noted at the beginning, it took almost 20 years, but by 1990 we had a fairly
solid estimate of the effect of inspections with penalties on lost workday injuries.
To be more precise, we had a fairly solid estimate of the average effect of such
inspections in federal OSHA states among manufacturing plants during the years
1979-85. This finding helped OSHA make a stronger analytical case for support of
its enforcement program. Otherwise, however, its practical impact has been
limited. Why?

A key problem is that studies are rarely designed to provide evidence on
current policy choices and, even when they are, the evidence they produce is often
not precise enough to provide much guidance. For example, even if we know that
the average effect of inspections is larger at smaller workplaces, we still will not
know what the incremental effect of a 10% or 50% shift would be. The average
effect may or may not be a good measure of the marginal effect of these changes.
Also, the effect may depend upon other characteristics of the workplace or the
inspection. Yet studies usually do not have sufficiently large samples to explore
many of these possible interactions.

These observations may suggest glum conclusions. And it is certainly true
that any one evaluation is likely to provide only limited insight. But this is true of
evaluations in all fields, even for areas like clinical trials of pharmaceuticals, which
represents perhaps the most developed use of scientific methods in practical affairs.
The implication is that what is needed is a large-scale and continuing commitment
to evaluation. This commitment must come from the regulatory agency; among
other reasons, only the agency can carry out the planned variations in prevention
activities that must be the foundation of any successful learning. An agency will
not generate many useful policy insights if it relies only upon retrospective data
analysis. It must consciously and continuously experiment with different
approaches and rigorously assess the results. The challenge to the agency should
be: "What have you learned this year about how to do your job more effectively?"

Such a policy requires that the agency have sufficient flexibility to be able to
experiment. Whether it does will depend partly upon leadership and the
organizational culture, but it will also depend upon the constraints imposed by the
political environment and by the law. The discretion that law enforcement agencies

can exercise will be confined by prevailing notions of statutory fidelity, due process, and equal protection. These can pose significant barriers. In the United States, two of the most promising initiatives in the last 15 years were overturned by the courts. In addition to derailing the Maine Top 200-inspired program described above, courts ruled that OSHA had failed to provide enough evidence to justify its 1989 Air Contaminants standard, which OSHA had projected would prevent more than 600 toxic substance deaths per year at a cost of about $1 million per death.[8] Overcoming these constraints will require, at a minimum, that organizational leadership make a strong commitment to learning. What would be truly depressing would be to come back in 10 or 20 years and find that the agency has learned little or nothing about how to improve its performance.

Unfortunately, the depressing scenario is probably the most likely. Bureaucracies tend to be defensive and not open to experimentation. Politicians tend to have little intrinsic interest in the outcomes of policies as long as key constituencies are satisfied. Business groups typically are more interested in limiting the costs of regulations than in increasing their effectiveness. Organized labor typically focuses on: 1) protecting the wages of those injured; and 2) ensuring that the enforcement agencies are responsive to their demands, but also keep pressure on non-union firms in order to prevent them from gaining a competitive advantage. Finding ways to become more effective at reasonable cost is not at the top of anyone's agenda.

11.2.9 Understanding What the Different Elements of Workplace Safety and Health Programs Contribute

Trade magazines are full of stories about "How My Firm Reduced Its Injury Rate by 70% in Just Six Months", but these stories rarely add to our knowledge base. However, a number of quantitative studies of the relation between establishment characteristics and injury rates have been carried out and there is enough agreement among them to allow identification of some factors that should be included in ratings of employer safety and health programs. For regulators, this information would potentially help to expand the scope of injury prevention beyond the set of injuries addressed by specific safety standards. The development of standards requiring that employers adopt "comprehensive safety and health programs" has been a recent priority in the United States at both the federal and state level. In addition, such information could provide an intermediate outcome measure that regulators could use to assess employer efforts and to assess their own success in inducing firms to improve their programs. Yet our understanding of the processes that lead to sustained good performance remains very inadequate. Even when we believe we know some factors that matter, we often lack useful ways to operationalize them and thus cannot measure them or measure their impact (Shannon, 2000). The result is that it is difficult for both regulators and the regulated to know what specific measures are worthwhile.

For an example of the operationalization issue, consider "management commitment to safety and health", which is often cited as the single most important influence on workplace safety. How should this be measured? One method could be to examine what role injury results play in the appraisals of supervisors and executives. Should this be measured on a "yes-no" scale or should it be more finely calibrated? How should this element be weighted in an overall measure of a safety program, compared to other factors? Should management commitment to safety be viewed as a zero-sum game, where more attention to safety and health means less attention to other matters, or as a marker of a well-managed firm?

OSHA's consultation program has used a method for rating employers' programs for a number of years; however, its validity and reliability had never been well-established. A revised instrument was developed which relies on "yes" or "no" answers to 81 questions. It was tested in approximately 240 workplaces (Smitheram and Weems, 2000). Workplaces were divided into high, medium, and low categories on both their injury rates and their scores. The instrument explained only 3% of the variation in injury rates. Apparently no effort was made to examine the relationship between individual answers and outcomes.

Although it would be desirable to have more answers to many of the questions raised here, it is probably preferable to develop and use some method, even one that is only roughly right. One advantage is that the data collected in this way could be used for further calibration of the instrument.

11.2.10 Understanding the Nature of the Occupational Disease Problem

We might think of the occupational disease problem as comprising several parts:

1 Diseases due to known hazards where current exposures exceed legal limits.
2 Diseases due to known hazards where exposures do not exceed legal limits.
3 Diseases due to hazards that are presently unknown or poorly understood.

Although there are overlaps, the first of these is primarily an enforcement problem; the second, primarily a standard-setting problem, and the third, primarily a problem in hazard recognition and assessment. All three need continuing attention and are subject to uncertainties, although the bounds on uncertainties are tighter for the first two. As mentioned for American data, overexposures to current OSHA standards are largely a problem for lead, silica, and noise.

In relation to new standards, OSHA estimated that approximately 600 deaths per year would be prevented by its 1989 Air Contaminants standard. For almost all of the almost 200 hazards covered by the standard, OSHA was proposing to reduce exposure levels to those recommended by the American Conference of Governmental Industrial Hygienists (ACGIH). Usually, these levels were 50% to 75% below the OSHA PELs. Therefore, even if OSHA went below the ACGIH limits for these substances and reduced its existing PELs by 90% or 95%, it would probably prevent another 400 or 500 deaths beyond the 600. These figures suggest that a reduction in many current PELs deserves immediate attention, although, given the morass of the rulemaking process in the United States, it is likely to be achieved only by legislative action. A problem deserving the most sustained attention is the one posed above, what more, if anything, needs to be done to ensure that new, potentially major occupational health hazards do not emerge? In other words, how do we prevent another "asbestos"?

The answer will depend upon many factors, including:

1 The extent of testing of new chemicals.
2 The extent of medical screening of workers beyond that mandated by substance-specific standards.
3 Whether new hazards cause diseases that are relatively uncommon outside of the workplace or diseases that are common.

The extent of medical screening may depend upon how we resolve what seems to be a dilemma facing employers. If they screen workers and discover a disease cluster or excess rate for some illness, what should employers do? Assume, that most, but not all, of such findings are false positives. Notifying the affected workers could alarm them and could lead to demands that the firm take action. On the other hand, not notifying them could lead, if the findings turned out to be true indications of a new hazard, to serious liability for the firm due to its failure to warn workers. It is easy to see how employers who would be faced with this choice might decide that the best solution was simply not to conduct medical screening. From a societal perspective, this outcome would be unfortunate.

The solution would appear to depend on developing some clear guidelines for employers about the conditions under which workers should be informed. In some settings, labor-management bargaining may be able to provide these guidelines. In other cases, we will have to rely either on tort law or on regulation. Because of the uncertainties and variation that affect tort law, a reliance on regulation may be desirable here (Dewees *et al.*, 1996).

11.3 KEY MESSAGES

- Government regulation of occupational safety and health conditions in workplaces has contributed to reductions in injuries and in toxic exposures. The impacts of new standards on toxic exposures have often been dramatic. Reductions in injury rates appear to have been much smaller. Most of the studies and most of the evidence for this conclusion come from the US.

- Generalizations from individual studies should be made cautiously; practices and conditions can change. Most notably, studies indicate that the average effects of OSHA inspections in manufacturing have declined since the first studies examined their effects in the early 1980s. The mechanisms underlying these changes are not well understood.

- The high rate of unionization and the high rate of inspections in Canada suggest that workplace safety and health programs there should be strong. Studies using aggregated Canadian data have found little impact of the effects of inspections, a finding that seems especially surprising in light of the limited degree of experience-rating in WC insurance. A careful comparison of both injury patterns and workplace practices in the US and Canada might help us to understand both countries better.

- Evidence about the limited current impact of inspections should spur efforts to design new strategies of intervention.

- More systematic learning about "what works" in regulation will probably require workplace regulatory agencies to take the task of learning more seriously. They are uniquely situated to systematically try out new strategies and to gather data.

11.4 NOTES

1. We also reviewed the literature on European regulation, but we found only one quantitative study of the effect of inspections on injuries. The UK Health and Safety Executive published Occasional Paper #11 in 1985 on "Measuring the Effectiveness of HSE's Field Activities". It looked at aggregate inspection rates and injury rates by industry and found that "neither visit frequency nor the number of notices and prosecutions had a significant effect on relative accident rates between industries in the same year". For a valuable discussion of occupational safety and health policy in the UK, see Nichols (1997). For recent descriptions of European programs, see "Priorities and Strategies in Occupational Safety and Health", a report by the European Union Agency for Safety and Health at Work, which can be found on their website at http://www.agency.osha.eu.int/reports/priorities. A great deal of descriptive material is also available from the European Foundation for the Improvement of Living and Working Conditions.

2. One aspect of this larger issue was raised in a study by Smith (1990) which found that MSDs and other injuries whose etiology is difficult to trace are disproportionately reported on Mondays, a finding that suggests that some may not be work-related. However, Card and McCall (1996) presented evidence indicating that the higher frequency of Monday injuries was not due to WC. Ruser (1998) found no evidence that "harder-to-diagnose" injuries occurred disproportionately on Mondays, but did link higher Monday injury rates to higher state WC payment levels.

3. Chamber of Commerce of the United States v. OSHA, D.C. Court of Appeals, Docket 98-1036, decided April 1999.

4. A somewhat more positive assessment of the impact of the Maine Top 200 program on injuries can be found in Marcus Stanley, "Essays in Program Evaluation", a PhD dissertation submitted to the Public Policy Program at Harvard University in 2000.

5. Personal communication, Bob Kusiak, Ontario Ministry of Labor.

6. Data for Ontario for 2000 were obtained from a personal communication with Ed McCloskey of the Ontario Ministry of Labor.

7. Enforcement in the United Kingdom seems to be intermediate between the US and Ontario, in several respects. The number of inspections was about 140,000 in 1995 for 56 million people, a rate about eight times higher than in the US, but well below Ontario's. As in Ontario, prosecutions in the UK were rare, but penalties there also appeared smaller than in Ontario's – an average of some $3,000 per prosecution compared to over $30,000 in Ontario (Nichols 1997).
8. AFL-CIO v. OSHA 965 F.2d 962 (11th Cir. 1992).

11.5 ACKNOWLEDGEMENTS

I would like to acknowledge useful comments on earlier drafts of this chapter and related papers from Wayne Gray, John Scholz, John Ruser, Terry Sullivan, and John Frank. Marianne Levitsky helped to steer me to the right people to talk to in the Ontario Ministry of Labour and Ed McCloskey and Bob Kusiak patiently answered a number of my questions. David Walters assisted in identifying references in the UK literature.

11.6 REFERENCES

Ayres, I. and Braithwaite, J., 1994, *Transcending Deregulation*, (New York: Oxford University Press).

Bartel, A.P. and Thomas, L.G., 1985, Direct and indirect effects of OSHA regulation: A new look at OSHA's impact. *Journal of Law and Economics*, **28**, pp. 1-25.

Bardach, E. and Kagan, R., 1983, *Going by the Book: The Problem of Regulatory Unreasonableness* (Philadelphia: Temple University Press).

Brown, R., 1994, Theory and practice of regulatory enforcement: occupational health and safety regulation in British Columbia. *Law and Policy*, **16**, pp. 63-91.

Bureau of Labor Statistics website (www.bls.gov) on data from 2000, *Census of Fatal Occupational Injuries*.

Card, D. and McCall, B.P., 1996, Is workers' compensation covering uninsured medical costs? Evidence from the 'Monday effect'. *Industrial and Labor Relations Review*, **49**, pp. 690-706.

Cousineau, J-M., Girard, S. and Lanoie, P., 1995, Safety regulation and specific injury types in Quebec. In *Research in Canadian Workers Compensation*, edited by Thomason, T. and Chaykowski, R.P. (IRC Press, Queen's University).

Dewees D, Duff, D. and Trebilcock, M., 1996, *Exploring the domain of accident law: taking the facts seriously*, (Oxford University Press).

Fine, L.J., Silverstein, B.A., Armstrong, T.J. and Anderson, C.A., 1984, An alternative way of detecting cumulative trauma disorders of the upper extremity in the workplace. In *Proceedings of the 1984 International Conference on Occupational Ergonomics*, Human Factors Association, Toronto, Canada. International Conference on Occupational Ergonomics, pp. 425-429.

Froines, J.R., Baron, S. Wegman, D.H., and O'Rourke, S., 1990, Characterization of the airborne concentration of lead in U.S. industry. *American Journal of Industrial Medicine*, **18**, pp .1-17.

Gomez, M.R., 1997, Factors associated with exposure in occupational safety and health administration data. *American Industrial Hygiene Association Journal*, **58**, pp. 1-10.

Gray, W.B. and Scholz, J.T., and 1990, OSHA enforcement and workplace injuries: A behavioral approach to risk assessment. *Journal of Risk and Uncertainty,* **3**, pp. 283-305.

Gray, W.B. and Scholz, J., 1993, Does regulatory enforcement work? A panel analysis of OSHA enforcement. *Law and Society Review*, **27**, pp. 177-213.

Gray, W.B. and Jones, C.A., 1991, Are OSHA health inspections effective? A longitudinal study in the manufacturing sector. *Journal of Human Resources*, **36**, pp. 623-653.

Gray, W.B., June 22, 1996, *Construction and analysis of BLS-OSHA matched data: Final report,* mimeo.

Gray, W.B. and Mendeloff, J.M., 2002, The declining effects of OSHA Inspections on Manufacturing Industries, 1979 to 1998, National Bureau of Economic Research Working Paper, No. W p. 9119.

Harrington, W., Morgenstern, R.D. and Nelson, P., 2000, On the accuracy of regulatory cost estimates. *Journal of Policy Analysis and Management*, **19**, pp. 733-754.

Heinzerling, L., 1998, Regulatory costs of mythic proportions. *Yale Law Journal* **107**, pp. 1981.

Kneisner, T.J. and Leeth, J.D., 1995, Numerical simulation as a complement to econometric research on workplace safety. *Journal of Risk and Uncertainty*, **10**, pp. 99-125.

Kneisner, T.J. and Leeth, J.D., 1989, Separating the reporting effects from the injury rate effects of workers compensation insurance: A hedonic simulation. *Industrial and Labor Relations Review*, **42**, pp. 280-293.

Kralj, B., 1998, Occupational health and safety: effectiveness of economic and regulatory mechanisms. In *Workers Compensation: Foundations for Reform*, edited by Gunderson, M. and Hyatt, D., (Toronto: University of Toronto Press).

Lanoie, P., 1992a, The impact of occupational safety and health regulation on the incidence of workplace accidents in Quebec 1983-87. *Journal of Human Resources,* **27**, pp. 643-660.

Lanoie, P., 1992b, Government intervention in occupational safety: Lessons from the American and Canadian experience. *Canadian Public Policy*, **18**, pp. 62-75.

Lanoie, P., 1992c, Safety regulation and the risk of workplace accidents in Quebec. *Southern Economic Journal*, **58**, pp. 950-965.

Lewchuk, W., Robb, A.L. and Walters V., 1996, The effectiveness of Bill 70 and joint health and safety committees in reducing injuries in the workplace: the case of Ontario. *Canadian Public Policy*, **22**, pp. 225-243.

Mendeloff, J.M., 1979, *Regulating Safety: An Economic and Poltical Analysis of Occupational Safety and Health Policy*, (Cambridge, MA: MIT Press).

Mendeloff, J.M., 1984, The role of OSHA violations in serious workplace accidents. *Journal of Occupational Medicine*, **26**, pp. 353-360.

Mendeloff, J.M., 1988, *The dilemma of toxic substance regulation: how overregulation causes underregulation at OSHA*, (Cambridge, MA: The MIT Press).

Mendeloff, J.M., 1996, *Issues Raised by the OSHA Data Initiative, A Report to the OSHA Office of Statistics*, September, 1996.

Moore, M.J. and Viscusi, W.K., 1990, *Compensating Mechanisms for Job Risk*, (Princeton, NJ: Princeton University Press).

Morrall, J.F., III., 1986, A review of the record. *Regulation*, Nov./Dec. 1986.

National Research Council, 1987, *Counting injuries and illnesses in the workplace*, (Washington, D.C.: National Academy Press).

Nichols, T., 1997, *The Sociology of Industrial Injury*, (London: Mansell Publishing).

O'Grady, J., 2000, Joint health and safety committees: finding a balance. In T. Sullivan (ed.) *Injury and the New World of Work*. (Vancouver: UBC Press), pp. 162-197.

Oleinick, A, Guire, K.E., Hawthorne, V.M, Guire, A., Hawthorne, V.M., Schork, M.A., Gluck, J.V., Lee, B. and Ha, S., 1993, Current Methods of estimating severity for occupational injuries and illnesses: Data from the 1986 Michigan comprehensive compensable injury and illness database. *American Journal of Industrial Medicine*, **23**, pp. 231-252.

Rivara, F.P. and Thomson, D.C., 2000, Systematic reviews of strategies to prevent occupational injuries, special issue. *American Journal of Preventive Medicine*, **18**, 4S.

Ruser, J.W., 1991, Workers' compensation and occupational injuries and illnesses. *Journal of Labor Economics*, **9**, pp. 325-350.

Ruser, J.W., 1993, Workers' compensation and the distribution of occupational injuries. *Journal of Human Resources*, **28**, pp. 593-617.

Ruser, J.W., 1998, Does workers' compensation encourage hard to diagnose injuries?. *Journal of Risk and Insurance*, **65**, pp. 101-124.

Ruser, J.W. and Smith, R.S., 1988, The effect of OSHA records-check inspections on reported occupational injuries in manufacturing. *Journal of Risk and Uncertainty*, **1**, pp. 415-435.

Ruser, J.W. and Smith, R.S., 1991, Re-estimating OSHA's Effects: Have the Data Changed? *Journal of Human Resources*, **26**, pp. 212-235.

Scholz, J.T. and Gray, W.B., 1997, Can government facilitate cooperation? An information model of OSHA enforcement. *American Journal of Political Science*, **41**, pp. 693-717.

Shannon, H., 2000, Firm-level organizational practices and work injuries. In *Injury and the New World of Work*, Chapter 7, edited by Sullivan, T. (UBC Press).

Shapiro, S.A. and Rabinowitz, R.S., 1997, Punishment versus cooperation in regulation enforcement: A case-study of OSHA. *Administrative Law Review*, **49**, pp. 713-762.

Siskind, F.B., 1993, Twenty years of OSHA federal enforcement data. *U.S. Department of Labor, Office of the Assistant Secretary for Policy,* mimeo, January 1993.

Smith, R.S., 1979, The impact of OSHA inspections on manufacturing injury rates. *Journal of Human Resources,* **14,** pp. 145-170.

Smith, R.S., 1986, Greasing the squeaky wheel: The relative productivity of OSHA complaint inspections. *Industrial and Labor Relations,* **10,** pp. 35-47.

Smith, R.S., 1990, Mostly on Mondays: is workers' compensation covering off-the-job injuries? In *Benefits, Costs and Cycles in Workers' Compensation,* edited by Borba, P.S. and Appel, D., (Boston: Kluwer Publishers).

Smith, R.S., 1992, Have OSHA and workers' compensation made the workplace safer. In *Research Frontiers in Industrial Relations and Human Resources,* edited by Lewin, D. and Mitchell, O.S., (Industrial Relations Research Association).

Smitherman, H., *et al.,* 2000, Consultation Evaluation Tool Prediction Report, mimeo.

Sullivan, T., 2000, *Injury and the New World of Work,* (UBC Press, Vancouver).

Tuohy, C. and Simard, M., 1992, The impact of joint health and safety committees in Ontario and Quebec. *Report to the Canadian Administrators of Labour Law,* (Toronto).

US Congress, Office of Technology Assessment, 1995, Gauging control technology and regulatory impacts in occupational safety and health: An appraisal of OSHA's analytic approach. *OTA-ENV-635* (Washington, D.C.: US Government Printing Office).

US Department of Labor, 1994, *Report to the House and Senate Appropriations Committees*: The Availability and Use of Data on Occupational Injuries and Illnesses, mimeo, July 1994.

Viscusi, W.K., 1983, *Risk by Choice: Regulating Health and Safety in the Workplace* (Cambridge, MA: Harvard University Press).

Viscusi, W. K., 1985, Cotton dust regulation: An OSHA success story? *Journal of Policy Analysis and Management,* **4,** pp. 325-343.

Viscusi, W. K., 1986, The impact of occupational safety and health regulation. *Rand Journal of Economics,* **17,** pp. 567-580.

Waehrer, G.M., Miller, T., Ruser, J. and Leigh, J.P., 1997, *Restricted Work, Workers' Compensation and Days Away from Work,* mimeo, December 1997.

Weil, D., 1996, If OSHA is so bad, why is compliance so good? *Rand Journal of Economics,* **27,** pp. 618-640.

Weil, D., 2000, *Assessing OSHA Performance: Evidence from the Construction Industry,* mimeo.

Workplace Health and Safety Agency (Ontario), 1994, *The Impact of Joint Health and Safety Committees on Health and Safety in Ontario,* (Toronto: Workplace Health and Safety Agency).

Index

Note: Figures and Tables are indicated by *italic page numbers,* notes by suffix "n" (e.g. "84n[1]" means note 1 on page 84). Abbreviations: LBP = low back pain; RSI = repetitive strain injuries; RTW = return to work; TRW = therapeutic return to work; WMSD = work-related musculoskeletal disorders